Wireless Communication for Cybersecurity

Scrivener Publishing
100 Cummings Center, Suite 541J
Beverly, MA 01915-6106

Advances in Antenna, Microwave, and Communication Engineering

Series Editor: Manoj Gupta, PhD, Pradeep Kumar, PhD

Scope: Antenna and microwave, as well as digital communication, engineering has been increasingly adopted in many diverse applications such as radio astronomy, long-distance communications, space navigation, radar systems, medical equipment's as well as missile electronic systems. As a result of the accelerating rate of growth of communication, microwave and antenna technology in research and industry sectors; students, teachers and practicing engineers in these area are faced with the need to understand various theoretical and experimental aspects of design and analysis of microwave circuits, antennas and simulation techniques, communication systems as well as their applications. Antennas, Microwave and Communication Engineering are actually a very lively and multidisciplinary one, mixing the deepest electromagnetic theoretical aspects, mathematical signal and data processing methods, physics of devices and physics of fields, software developments and technological fabrication aspects, and the large number of possible applications generates multiple outcomes. Hence the aim of this book series is to provide a multi-discipline forum for engineers and scientists, students, researchers, industry professionals in the fields of Antenna, Microwave, Communication and Electromagnetic Engineering to focus on advances and applications.

Publishers at Scrivener
Martin Scrivener (martin@scrivenerpublishing.com)
Phillip Carmical (pcarmical@scrivenerpublishing.com)

Wireless Communication for Cybersecurity

Edited by

S. Sountharrajan
R. Maheswar
Geetanjali Rathee
and
M. Akila

Scrivener Publishing

WILEY

This edition first published 2023 by John Wiley & Sons, Inc., 111 River Street, Hoboken, NJ 07030, USA and Scrivener Publishing LLC, 100 Cummings Center, Suite 541J, Beverly, MA 01915, USA
© 2023 Scrivener Publishing LLC
For more information about Scrivener publications please visit www.scrivenerpublishing.com.

All rights reserved. No part of this publication may be reproduced, stored in a retrieval system, or transmitted, in any form or by any means, electronic, mechanical, photocopying, recording, or otherwise, except as permitted by law. Advice on how to obtain permission to reuse material from this title is available at http://www.wiley.com/go/permissions.

Wiley Global Headquarters
111 River Street, Hoboken, NJ 07030, USA

For details of our global editorial offices, customer services, and more information about Wiley products visit us at www.wiley.com.

Limit of Liability/Disclaimer of Warranty
While the publisher and authors have used their best efforts in preparing this work, they make no representations or warranties with respect to the accuracy or completeness of the contents of this work and specifically disclaim all warranties, including without limitation any implied warranties of merchantability or fitness for a particular purpose. No warranty may be created or extended by sales representatives, written sales materials, or promotional statements for this work. The fact that an organization, website, or product is referred to in this work as a citation and/or potential source of further information does not mean that the publisher and authors endorse the information or services the organization, website, or product may provide or recommendations it may make. This work is sold with the understanding that the publisher is not engaged in rendering professional services. The advice and strategies contained herein may not be suitable for your situation. You should consult with a specialist where appropriate. Neither the publisher nor authors shall be liable for any loss of profit or any other commercial damages, including but not limited to special, incidental, consequential, or other damages. Further, readers should be aware that websites listed in this work may have changed or disappeared between when this work was written and when it is read.

Library of Congress Cataloging-in-Publication Data

ISBN 9781119910435

Front cover images supplied by Pixabay.com
Cover design by Russell Richardson

Set in size of 11pt and Minion Pro by Manila Typesetting Company, Makati, Philippines

Printed in the USA

10 9 8 7 6 5 4 3 2 1

Contents

Preface xiii

1 BBUCAF: A Biometric-Based User Clustering Authentication Framework in Wireless Sensor Network 1
Rinesh, S., Thamaraiselvi, K., Mahdi Ismael Omar and Abdulfetah Abdulahi Ahmed
   1.1 Introduction to Wireless Sensor Network 2
   1.2 Background Study 3
   1.3 A Biometric-Based User Clustering Authentication Framework 5
      1.3.1 Biometric-Based Model 5
      1.3.2 Clustering 9
   1.4 Experimental Analysis 12
   1.5 Conclusion 16
   References 16

2 DeepNet: Dynamic Detection of Malwares Using Deep Learning Techniques 21
Nivaashini, M., Soundariya, R. S., Vishnupriya, B. and Tharsanee, R. M.
   2.1 Introduction 22
   2.2 Literature Survey 24
      2.2.1 ML or Metaheuristic Methods for Malware Detection 25
      2.2.2 Deep Learning Algorithms for Malware Detection 26
   2.3 Malware Datasets 28
      2.3.1 Android Malware Dataset 28
      2.3.2 SOREL-20M Dataset 28
   2.4 Deep Learning Architecture 29
      2.4.1 Deep Neural Networks (DNN) 29
      2.4.2 Convolutional Neural Networks (CNN) 29
      2.4.3 Recurrent Neural Networks (RNN) 30

		2.4.4 Deep Belief Networks (DBN)	30
		2.4.5 Stacked Autoencoders (SAE)	31
	2.5	Proposed System	32
		2.5.1 Datasets Used	32
		2.5.2 System Architecture	34
		2.5.3 Data Preprocessing	35
		2.5.4 Proposed Methodology	35
		2.5.5 DeepNet	37
		2.5.6 DBN	37
		2.5.7 SAE	38
		2.5.8 Categorisation	39
	2.6	Result and Analysis	40
	2.7	Conclusion & Future Work	51
		References	51
3	**State of Art of Security and Risk in Wireless Environment Along with Healthcare Case Study**		**55**
	Deepa Arora and Oshin Sharma		
	3.1	Introduction	56
	3.2	Literature Survey	58
	3.3	Applications of Wireless Networks	60
	3.4	Types of Attacks	62
		3.4.1 Passive Attacks	62
		3.4.2 Release of Message Contents	62
		3.4.3 Traffic Analysis	63
		3.4.4 Eavesdropping	63
	3.5	Active Attacks	63
		3.5.1 Malware	64
		3.5.2 Password Theft	64
		3.5.3 Bandwidth Stealing	64
		3.5.4 Phishing Attacks	64
		3.5.5 DDoS	64
		3.5.6 Cross-Site Attack	65
		3.5.7 Ransomware	65
		3.5.8 Message Modification	65
		3.5.9 Message Replay	66
		3.5.10 Masquerade	66
	3.6	Layered Attacks in WSN	66
		3.6.1 Attacks in Physical Layer	67
		3.6.2 Attacks in Data Link Layer	67
		3.6.3 Attacks in Network Layer	68

		3.6.4	Attacks in Transport Layer	68
		3.6.5	Attacks in Application Layer	69
	3.7	Security Models		69
		3.7.1	Bio-Inspired Trust and Reputation Model	69
		3.7.2	Peer Trust System	70
	3.8	Case Study: Healthcare		71
		3.8.1	Security Risks in Healthcare	72
		3.8.2	Prevention from Security Attacks in Healthcare	73
	3.9	Minimize the Risks in a Wireless Environment		74
		3.9.1	Generate Strong Passwords	74
		3.9.2	Change Default Wi-Fi Username and Password	75
		3.9.3	Use Updated Antivirus	75
		3.9.4	Send Confidential Files with Passwords	75
		3.9.5	Detect the Intruders	75
		3.9.6	Encrypt Network	76
		3.9.7	Avoid Sharing Files Through Public Wi-Fi	76
		3.9.8	Provide Access to Authorized Users	76
		3.9.9	Used a Wireless Controller	76
	3.10	Conclusion		76
		References		77
4	**Machine Learning-Based Malicious Threat Detection and Security Analysis on Software-Defined Networking for Industry 4.0**			**79**
	J. Ramprasath, N. Praveen Sundra Kumar, N. Krishnaraj and M. Gomathi			
	4.1	Introduction		80
		4.1.1	Software-Defined Network	80
		4.1.2	Types of Attacks	81
			4.1.2.1 Denial of Services	81
			4.1.2.2 Distributed Denial of Service	84
	4.2	Related Works		86
	4.3	Proposed Work for Threat Detection and Security Analysis		89
		4.3.1	Traffic Collection	89
			4.3.1.1 Data Flow Monitoring and Data Collection	89
			4.3.1.2 Purpose of Data Flow Monitoring and Data Collection	90
			4.3.1.3 Types of Collection	91
		4.3.2	Feature Selection Using Entropy	93
		4.3.3	Malicious Traffic Detection	94
			4.3.3.1 Framing of the Expected Traffic Status	95

		4.3.3.2 Traffic Filtering Using Regression	95
	4.3.4	Traffic Mitigation	95
4.4	Implementation and Results		96
4.5	Conclusion		100
	References		101

5 **Privacy Enhancement for Wireless Sensor Networks and the Internet of Things Based on Cryptological Techniques**   **105**
Karthiga, M., Indirani, A., Sankarananth, S., S. S. Sountharrajan and E. Suganya

5.1	Introduction	106
5.2	System Architecture	107
5.3	Literature Review	108
5.4	Proposed Methodology	112
5.5	Results and Discussion	118
5.6	Analysis of Various Security and Assaults	122
5.7	Conclusion	124
	References	124

6 **Security and Confidentiality Concerns in Blockchain Technology: A Review**   **129**
G. Prabu Kanna, Abinash M.J., Yogesh Kumar, Jagadeesh Kumar and E. Suganya

6.1	Introduction		130
6.2	Blockchain Technology		131
6.3	Blockchain Revolution Drivers		133
	6.3.1	Transparent, Decentralised Consensus	133
	6.3.2	Model of Agreement(s)	134
	6.3.3	Immutability and Security	134
	6.3.4	Anonymity and Automation	134
	6.3.5	Impact on Business, Regulation, and Services	135
	6.3.6	Access and Identity	135
6.4	Blockchain Classification		135
	6.4.1	Public Blockchain	136
	6.4.2	Private Blockchain	137
	6.4.3	Blockchain Consortium	137
6.5	Blockchain Components and Operation		138
	6.5.1	Data	139
	6.5.2	Hash	139
	6.5.3	MD5	139
	6.5.4	SHA 256	140

		6.5.5	MD5 vs. SHA-256	140
	6.6	Blockchain Technology Applications		142
		6.6.1	Blockchain Technology in the Healthcare Industry	142
		6.6.2	Stock Market Uses of Blockchain Technology	142
		6.6.3	Financial Exchanges in Blockchain Technology	143
		6.6.4	Blockchain in Real Estate	143
		6.6.5	Blockchain in Government	144
		6.6.6	Other Opportunities in the Industry	145
	6.7	Difficulties		145
	6.8	Conclusion		145
		References		145
7	**Explainable Artificial Intelligence for Cybersecurity**			**149**
	P. Sharon Femi, K. Ashwini, A. Kala and V. Rajalakshmi			
	7.1	Introduction		150
		7.1.1	Use of AI in Cybersecurity	150
		7.1.2	Limitations of AI	151
		7.1.3	Motivation to Integrate XAI to Cybersecurity	151
		7.1.4	Contributions	152
	7.2	Cyberattacks		152
		7.2.1	Phishing Attack	153
			7.2.1.1 Spear Phishing	153
			7.2.1.2 Whaling	153
			7.2.1.3 Smishing	153
			7.2.1.4 Pharming	153
		7.2.2	Man-in-the-Middle (MITM) Attack	154
			7.2.2.1 ARP Spoofing	154
			7.2.2.2 DNS Spoofing	154
			7.2.2.3 HTTPS Spoofing	154
			7.2.2.4 Wi-Fi Eavesdropping	154
			7.2.2.5 Session Hijacking	155
		7.2.3	Malware Attack	155
			7.2.3.1 Ransomware	155
			7.2.3.2 Spyware	155
			7.2.3.3 Botnet	156
			7.2.3.4 Fileless Malware	156
		7.2.4	Denial-of-Service Attack	156
		7.2.5	Zero-Day Exploit	156
		7.2.6	SQL Injection	156
	7.3	XAI and Its Categorization		157
		7.3.1	Intrinsic or Post-Hoc	158

		7.3.2	Model-Specific or Model-Agnostic	159
		7.3.3	Local or Global	159
		7.3.4	Explanation Output	159
	7.4	XAI Framework		160
		7.4.1	SHAP (SHAPley Additive Explanations) and SHAPley Values	160
			7.4.1.1 Computing SHAPley Values	161
		7.4.2	LIME - Local Interpretable Model Agnostic Explanations	162
			7.4.2.1 Working of LIME	163
		7.4.3	ELI5	163
		7.4.4	Skater	164
		7.4.5	DALEX	164
	7.5	Applications of XAI in Cybersecurity		165
		7.5.1	Smart Healthcare	166
		7.5.2	Smart Banking	166
		7.5.3	Smart Cities	166
		7.5.4	Smart Agriculture	167
		7.5.5	Transportation	167
		7.5.6	Governance	168
		7.5.7	Industry 4.0	168
		7.5.8	5G and Beyond Technologies	169
	7.6	Challenges of XAI Applications in Cybersecurity		169
		7.6.1	Datasets	170
		7.6.2	Evaluation	170
		7.6.3	Cyber Threats Faced by XAI Models	170
		7.6.4	Privacy and Ethical Issues	171
	7.7	Future Research Directions		171
	7.8	Conclusion		171
		References		172
8	**AI-Enabled Threat Detection and Security Analysis**			**175**
	A. Saran Kumar, S. Priyanka, V. Praveen and G. Sivapriya			
	8.1	Introduction		176
		8.1.1	Phishing	176
		8.1.2	Features	178
		8.1.3	Optimizer Types	179
		8.1.4	Gradient Descent	180
		8.1.5	Types of Phishing Attack Detection	181
	8.2	Literature Survey		181
	8.3	Proposed Work		184

		8.3.1	Data Collection and Pre-Processing	186
		8.3.2	Dataset Description	188
		8.3.3	Performance Metrics	189
	8.4	System Evaluation		190
	8.5	Conclusion		195
		References		195

9 **Security Risks and Its Preservation Mechanism Using Dynamic Trusted Scheme** 199
Geetanjali Rathee, Akshay Kumar, S. Karthikeyan and N. Yuvaraj

	9.1	Introduction		200
		9.1.1	Need of Trust	200
		9.1.2	Need of Trust-Based Mechanism in IoT Devices	201
		9.1.3	Contribution	201
	9.2	Related Work		202
	9.3	Proposed Framework		205
		9.3.1	Dynamic Trust Updation Model	205
		9.3.2	Blockchain Network	207
	9.4	Performance Analysis		209
		9.4.1	Dataset Description and Simulation Settings	209
		9.4.2	Traditional Method and Evaluation Metrics	210
	9.5	Results Discussion		210
	9.6	Empirical Analysis		212
	9.7	Conclusion		213
		References		213

10 **6G Systems in Secure Data Transmission** 217
A.V.R. Mayuri, Jyoti Chauhan, Abhinav Gadgil, Om Rajani and Soumya Rajadhyaksha

	10.1	Introduction			218
	10.2	Evolution of 6G			219
	10.3	Functionality			222
		10.3.1	Security and Privacy Issues		223
			10.3.1.1	Artificial Intelligence (AI)	223
			10.3.1.2	Molecular Communication	225
			10.3.1.3	Quantum Communication	226
		10.3.2	Blockchain		226
		10.3.3	TeraHertz Technology		227
		10.3.4	Visible Light Communication (VLC)		228
	10.4	6G Security Architectural Requirements			230

	10.5	Future Enhancements	234
	10.6	Summary	237
		References	237
11	**A Trust-Based Information Forwarding Mechanism for IoT Systems**		**239**
	Geetanjali Rathee, Hemraj Saini, R. Maheswar and M. Akila		
	11.1	Introduction	240
		11.1.1 Need of Security	240
		11.1.2 Role of Trust-Based Mechanism in IoT Systems	240
		11.1.3 Contribution	243
	11.2	Related Works	243
	11.3	Estimated Trusted Model	247
	11.4	Blockchain Network	248
	11.5	Performance Analysis	250
		11.5.1 Dataset Description and Simulation Settings	251
		11.5.2 Comparison Methods and Evaluation Metrics	251
	11.6	Results Discussion	252
	11.7	Empirical Analysis	253
	11.8	Conclusion	255
		References	255
About the Editors			**259**
Index			**261**

Preface

Wireless communication has become essential for everyday life all over the world, in almost every country. Irrespective of place or situation, people depend on wireless communication to fulfil their necessities. It is nearly impossible to remember a world before wireless communication became a critical entity in billions of lives. Rapid advancement in wireless communications and related technologies has led to advances in this domain, which is the use of newer technologies like 6G, IoT, radar, etc. Not only are these technologies expanding, but the impact of wireless communication is also changing and becoming an inevitable part of our lives.

With use comes responsibility with a lot of disadvantages for any newer technology. The growing risks in terms of security, authentications, user privacy, and encryptions are some major areas of concern. We have seen significant development in blockchain technology along with development in a wireless network that has proved extremely useful in solving many security issues. An efficient secure cyber-physical system can be constructed using these technologies. This book covers all kinds of situations regarding the digital health and processes of intrusion detection in wireless networks. It allows the readers to reach their solutions using various predictive algorithm-based approaches and some curated real-time protective examples that are defined. The chapters also comprehensively state the challenges in privacy and security levels for various algorithms and various techniques and tools are proposed for each challenge.

It focuses on exposing readers to advances in data security and privacy of wider domains. Security vulnerabilities are overcome using the techniques as proposed in the chapters. The book aims to address all viable solutions to the various problems faced in the newer techniques of wireless communications, improving the accuracies and reliability over the possible vulnerabilities and security threats to wireless communications. This book is useful for the researchers, academicians, R&D organizations, and healthcare professionals working in the area of antenna, 5G/6G communication, wireless communication, digital hospital, and intelligent medicine.

The key features of the book are:

- Serves as a strong technological convergence solution for wireless communications in the cyber security domain
- Enlightens the foundation of wireless communication networks embedding with cyber-physical systems and foundational topics of blockchain
- Exploring the practical issues in the automation domain
- Highlights the AI powered analytics to analyse the characteristics of wireless user behaviour security models
- Key insights about blockchain joining forces with wireless communication security to set up flawless cyber-physical systems

Dr. S. Sountharrajan
Dr. R. Maheswar
Dr. Geetanjali Rathee
Dr. M. Akila
September 2023

1
BBUCAF: A Biometric-Based User Clustering Authentication Framework in Wireless Sensor Network

Rinesh, S.[1*], Thamaraiselvi, K.[2], Mahdi Ismael Omar[1] and Abdulfetah Abdulahi Ahmed[1]

[1]Department of Computer Science, Jigjiga University, Jijiga, Ethiopia
[2]Department of Computer Science, Malla Reddy College of Engineering, Hyderabad, Telangana, India

Abstract

Wireless Sensor Networks (WSN) have made much progress in the last few years, so data transmission must be more secure. Cryptographic keys keep information private, authenticate people, and keep data safe. Several research projects were done to interact with important management issues in WSNs. Prime statistics are used to make collective keys. It would then be able to accurately check the security of nodes. A new network way is modeled for sending data between nodes without restriction. A strong authentication system is needed to maintain network safety and allow people to use a network service freely. But the limited supplies of sensor nodes make it tough to authenticate people. To overcome the security-based issues, a biometric-based user clustering authentication framework (BBUCAF) has been introduced to increase the level of security and the network's speed among the nodes. A biometric-based model is created by taking features from the fingerprint. Securely, feature vectors create a private key for the user. Such a key is sent to every sensor node. Then, private keys between sensor nodes are made by combining a randomly generated count and the user's key, which is sent to each sensor node. C- means Clustering is used to group nodes based on their range and unique identification. A collective key is made here using a fuzzy registration component that considers prime numbers. Fuzzy membership and biometric-based secret keys send data between groups and sensor nodes. Each cluster has group keys that differ from one cluster to the next. The network's speed improves the network's

---

Corresponding author: rin.iimmba@gmail.com

S. Sountharrajan, R. Maheswar, Geetanjali Rathee, and M. Akila (eds.) Wireless Communication for Cybersecurity, (1–20) © 2023 Scrivener Publishing LLC

effectiveness by cutting down on network traffic, protecting against DoS attacks, and extending the battery capacity of a node's battery with less energy usage.

Keywords: Wireless sensor network, nodes, clustering, network traffic, authentication

1.1 Introduction to Wireless Sensor Network

Several sensors can be used together in a single WSN. Nodes in the network that sense their surroundings are known as sensor nodes [1]. A wide range of applications, such as structural health monitoring, environmental control, and combat observation, can benefit from such connections [2]. A node can perform computing, identify itself, and communicate with other devices [3]. Those nodes can be dispersed in a situation where they can identify each other and work together to accomplish the task in a large region [4, 5]. Sensor nodes in WSNs are used for specific tasks [6]. Small sensor nodes in the network model their surroundings' information after spotting it [7]. Due to their wide range of applications, WSNs are becoming increasingly popular in education and the market [8]. WSNs are primarily designed to gather and send environmental information to a home or remote location via a network of sensing devices located in an isolated community [9]. The original data are then processed online or offline as per application standards for a full evaluation in a remote location [10]. If a patient is not in the hospital, for example, remote patient tracking is important for doctors.

These systems can benefit from numerous applications, including structural health monitoring, environmental control, and combat monitoring [11]. Most apps allow users to obtain data immediately from a gateway node because queries are handled on this node in most cases [12]. The information from a gateway node is very hard to receive on rare occasions. Therefore, sensor nodes collect information directly [13]. By sending the request to a sensor node, unauthorized users can quickly obtain sensitive information [14]. As a result of sensor nodes' inability to verify query messages may leak sensitive data, and network resources, such as node power and bandwidth, could be wastefully depleted [15]. Any or all of the associated issues could impact the network's lifespan and effectiveness, making the system inaccessible to genuine people [16]. Since network data and resources can be illegally accessed, authentication is necessary [17]. To achieve this, sensor nodes must validate users' identities [18]. All of the following issues can be solved with user authentication, which enables authorized users to join a system [19]. As a result of the resource limits of WSNs' small sensor devices, namely their power and storage, along with their processing and transmission capabilities, providing authentication in these networks is a very difficult issue [20]. Even though several standards have been presented, the authentication procedure is still vulnerable. In the end, a more robust and intelligent process is needed to assure the security of a WSN [21]. Maintaining a safe network requires a robust authentication system that allows users to access network services without restriction. Authentication is difficult due to the restricted supply of sensor nodes [22,23]. To overcome all the above-mentioned security-based issues, BBUCAF has been developed. The main contribution of BBUCAF is

- ➤ To build a biometric model, enhance the network's security and performance using fingerprints' unique characteristics.
- ➤ The user's private key is generated securely using feature vectors. Every sensor node receives a key. Then, each sensor node receives a random count, and the user's private key is combined with each sensor node.
- ➤ Numerous benefits of a faster network include reducing network traffic, preventing denial-of-service (DDoS) assaults, and increasing node battery life.

1.2 Background Study

Many researchers have carried out research works. Tsu-Yang Wu *et al.* [24] developed Three-Factor Authentication Protocol (TAP), in which the logical study and informal analysis confirm safety, Burross-Abadii-Needham (BAN) logic, and ProVerif tools. The evaluation of security and performance reveals that the method offers stronger security and reduced computational burden.

P.P. Devi *et al.* [25] proposed SDN-Enabled Hybrid Clone Node Detection Mechanisms (SDN-HCN). An SDN-based methodology performs a network path evaluation and time-based research methodologies to identify and reduce duplicate nodes produced by cloning attacks. To identify clone nodes in a wireless network, one must use the HCN technique. The simulation results reveal that several metrics are analyzed in the experiment.

M. Rakesh Kumar *et al.* [26] introduced a Secure Fuzzy Extractor-based Biometric Key Authentication (SFE-BKA) Scheme. The hash function is critical to the system's security. In SFE-BKA, the hash parameter value is irrespective of hash functions in an attempt to improve information security. The proposed method is not affected by this variance in hashing in terms of latency or delay. The outcome of SFE-BKA yielded 40% less data loss, 20% less energy usage, and less latency than earlier encoding systems.

S. Ashraf *et al.* [27] developed a Depuration-based Efficient Coverage Mechanism (DECM). Two rounds of deployment are required to complete the process. When a node is to be moved to new locations, the Dissimilitude Enhancement Scheme (DES) is used to find it. The Depuration mechanism in the second cycle reduces the separation between prior and new places by controlling the needless migration of the sensor nodes. By analyzing the simulation findings and computing in 0.016 seconds, the DECM has attained more than 98% protection.

Fan Wu *et al.* [28] described the Authentication Protocol for Wireless Sensor Networks (AP-WSN). Proverif's formal verification shows that the new system retains its security features. AP-WSN is feasible and meets general demands in a way that counters various threats and meets security properties. The proposed approach outperforms previous schemes in terms of security and is suitable for use. The simulation findings indicate that the plan may be successfully implemented in an IoT system and have a practical use.

Diksha Rangwani *et al.* [29] discussed improved privacy-preserving remote user authentication (PP-RUA). The suggested system is formally analyzed using the probabilistic Random-Oracle-Model to show the resilience of the scheme. Further, the system is simulated using a well-accepted AVISPA tool to show its security strength. The performance assessment of the system demonstrates that along with its consistency in aspects of privacy, the suggested scheme is more effective in computing and networking overheads than other current schemes.

SungJin Yu *et al.* [30] discussed Secure and Lightweight Three-Factor-Based User Authentication (SLUA). Secure, untraceable, and mutually authenticated communications are possible with the SLUA. Informal and formal methods are used to assess the safety of SLUA, along with the logic of Burrows–Abadi–Needham (BAN), the Real-or-Random (ROR) model, and the AVISPA simulation. SLUA-performance of WSNs is compared to other existing systems. Security and efficiency are more protected and more efficient in the proposed SLUA than in the prior suggested technique.

More problems are associated with security-based problems in sensor networks, and such security issues are concentrated on the proposed BBUCAF, and the obtained experimental analysis is compared with [25], [26], [27].

1.3 A Biometric-Based User Clustering Authentication Framework

1.3.1 Biometric-Based Model

The biometric-based model starts with the Registration Stage; a verified node serves as the initial registration point for new users. The biometric features of the users are then captured, and a hash is calculated based on the features. The entire architecture of BBUCAF is shown in Figure 1.1. From the biometric-based model, the features are extracted by transform, and based on the node identification, C-means clustering is implemented. The fuzzy membership function sends the data between the cluster and the sensor node. The effectiveness of the network is increased in terms of less traffic among the networks and protection against attacks.

The verified network would then receive their identification and hash code, as shown below

$$\left.\begin{array}{l} a = [identity_x, y] \\ y = g(biometrix.x) \end{array}\right\} \quad (1.1)$$

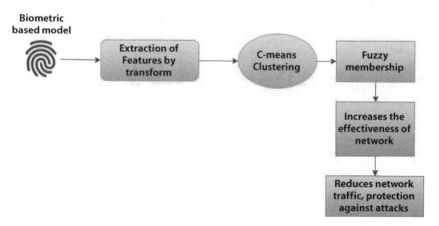

Figure 1.1 Architecture of BBUCAF.

The identification *a* and the hash code *y* are obtained from Equation (1.1), *identity$_x$* denote the identity element, *x*, *y* represent the verification node, *g* denotes the registration phase. The verified node calculates r after valid registration and transmits it to the user, as indicated in Equation (1.2). Verified nodes use the registered value to obtain the necessary information from the network.

$$\left. \begin{array}{l} a_o = [r] \\ r = g \left[\dfrac{identity_x}{u_0} \right] \end{array} \right\} \quad (1.2)$$

The valid registration r is obtained from Equation (1.2), and the next stage of identification is represented as a_o, *identity$_x$* denote the identity element, *g* denotes the registration phase, u_0 denote the initial stage. The different stages to getting the encrypted data are shown in Figure 1.2. The biometric identification phase captures the user's biometric data, and the verification stage compares the collected data to the stored data. The registration phase compiles new data to be given in the verification stage, and all the data collected from the sensor node is given to the registration phase. The biometric data and the hash values are compared to overcome all the attacks.

Biometric data is captured and hashed a second-time data is being sent to the sensor node together with *identity$_x$* and the required information

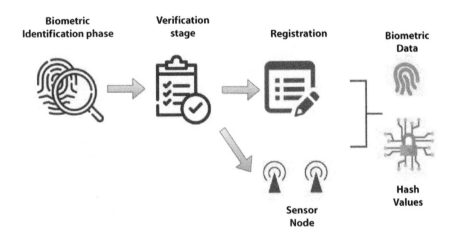

Figure 1.2 Different stages to getting the encrypted data.

is given in Equation (1.3). $y' = g(biometric)$ represent a user's most recent biometric information.

$$y' = g(biometric, RI, t_0) \qquad (1.3)$$

Here t_0 represent the user's actual period, RI represents the required information; initially, the sensor network examines the time stamp of a message it gets at the time t_1. The query is refused if $t_1 - t_0 < \Delta t$, else the demand is routed to a known network at a period t_2 for user authentication using its own identity. Δt represent the estimated time; the sensor network is used to identify the sensor node capable of responding to user inquiries.

$$\begin{cases} a_1 = [identity_x, b, t_2] \\ b = g[identity_x, (y')] \end{cases} \qquad (1.4)$$

The responding stage for user queries a_1 with the sensor network to reduce network traffic are obtained from Equation (1.4), b represents the user authentication, $identity_x$ denote the identity element, t_2 represent the period of the known network. y' represent a user's most recent biometric information. The verified node checks y and y', then the trustworthy node delivers a decline signal to the sensor node. $a_2 = [decline]$, the signal is sent to the client by the sensor network. $a_3 = [decline]$, the transmitter sends the data to a_4. $a_4 = [going\ on]$ After receiving a notification with the label for going on the process, the customer can begin the verification procedure.

The authentication stage and secret key generation for each cluster head are shown in Figure 1.3. The user authentication stage involves the extraction of features from the biometric sign, and the secret key generation is used for different sensor nodes m_x and m_y in the cluster group. Each network device has a clear text and an encrypted random counterpart as part of the configuration process.

After the authentication step, the feature patterns are taken from the fingerprint. The feature patterns are taken from the user where the secret keys are created. Each sensor node receives a copy of such a key using a pseudorandom character generator. From one cluster to the next, there are unique cluster group keys. As each node has access to both private and public data, it is possible for adjacent nodes to silently share a key. Authentication keys are used in the next phase to ensure that each pair of nearby nodes has a unique key before beginning a secure connection via an authorized connection. As a result of this strategy, nodes m_x and m_y are

8 Wireless Communication for Cybersecurity

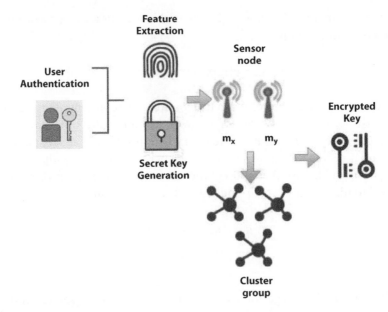

Figure 1.3 The authentication stage and secret key generation for each cluster head.

protected from each other by securely distributing comparable values. b_{ab} A pairwise biometric key uses pseudorandomness and the biometric feature vector to provide a secure result. Each network device has a clear text and an encrypted random counterpart (q_a, D_a) as part of the preconfiguration process. In addition, a small number of randomly selected prime numbers are divided into groups, with each junction of two different groups containing a single positive number.

The goal of such groups is to divide up the work of computing keys between each pair of adjacent nodes. Every node, m_x and m_y selects one cluster at random, and the intersection significance of the two groups is taken to be an overall prime number selected by m_x and m_y. A biometric cost and a pseudorandom feature are used to make b_{ab}. These are then thought to be hidden data and must be sent safely. After extraction, a hacker can immediately target any node to get back the used expert key. So, in this case, la_u is just added to certain nodes. Before the implementation stage, add $l_{to}l$ nodes to make it less likely that an attacker would be able to retrieve a volatile key without taking away from its short lifespan. Because each node already has private and public data, it needs to transfer the biometric couple key; this goal can be achieved before installation begins.

If la_u is placed into networks shielded from attackers, this strategy works well. lu can be incorporated into barrier sensors through an attack.

Even though an opponent manipulates and gives up the cluster from the network, the probability of obtaining a node containing a special key is la/M, where la is the number of networks with expert keys.

A random encrypted key is generated for the user XR_i when he or she registers with the system. The generated key would be stored on the Access point as a key for XR_i. A Biometric Encrypted pattern is created by user key with connection nodes and extracting the properties of the user identity from the transform. Random encrypted is used for validation and to generate a key from the fingerprint of the individual from XR_i. When using a piece of the fingerprint pattern for unencrypted biometric encoding, the pattern is massive; a pseudorandom function was used to generate results of variable size from inputs of a set size. In addition, each node has a random value saved in the Biometric Encryption template that can decline invalid keys at the beginning of the authentication process.

The wavelet transform converts a physical domain to a spectral domain. It is a common misconception that wavelet transforms on a pixel-by-pixel basis separate low- and high-frequency information. First do a one-dimensional transformation on each row, and then do the same for each column in a two-dimensional change.

1.3.2 Clustering

Clustering is based on the construction process for groups. In the beginning, a hijacked node can present information to a node's portion in a cluster while avoiding distribution to other devices in the network. The network structure separates a cluster into multiple sub-clusters, which minimizes the cluster's bulk data. The number of nodes in a cluster affects the probability that a corrupted node would be selected as the Clustering Head by chance. Let's imagine that there is indeed a group with a few participants and a future cluster with additional participants, each of which has damaged nodes. A team leader and susceptible nodes can help with the selection process. In these cases, a cluster with a damaged node as a group head is anticipated. The grouping key is distributed throughout collections for group-to-group interaction. Each device in the network receives a biometric feature array from the cluster head. The data collected from the biometric feature array is divided into N clusters, and each data item is partially assigned to each group using the c-means data clustering method. As an illustration, a data view's degree of participation in a collection increases the closer it is to its center. In contrast, a data point's degree of participation decreases the further it is from the cluster's center. C-means clustering is executed using the M_c Function. The starting point is a random estimate where the cluster centers and the average position of each cluster are determined. Following this, M_c gives each data point a random membership grade in each cluster. M_c realigns the cluster centers within a data set and calculates the level of participation in each cluster for every data point by dynamically adjusting the cluster centers and the involvement grades. To begin cluster construction, a cluster head node emits a clustering signal. Biometric characteristic vectors and the signal type are verified using private keys to prevent malicious activities. In Figure 1.4, x_{b1}, x_{b2}, x_{b3} …. x_{bn} represent the level of consistency of different users, and their biometric characteristics are fed to group keying in which the various DOS attacks are prevented in the sensor network.

If a node receives information from several cluster member networks, the recipient rejects the other information and links to the first one that arrives. Nodes verify their keys if they get a cluster member message. It joins the group and transmits the signal if it is verified correctly. The cluster member messages provide information on a cluster's primary key. Sending a signal simply adds the signal type and cluster message node's identity to the statement. The cluster head and its members interact with the sink node in the cluster head. The information channel between the cluster members and the group head is exposed.

Let $= \{x_a\}, a = 1, \ldots M_c$, M_c is the number of nodes in the Sensor Network configuration specified in the cluster. As n-dimensional vectors are represented as $\emptyset = \{\alpha_b \epsilon A^I, b = 1, \ldots a\}$ and reflect the clusters generated by nodes in A. For the sake of network configuration, let the vector is represented as $M * n$ with $X_{a,b}$ where a,b indicates the level of consistency of A_a with the

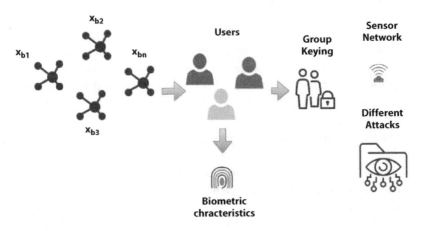

Figure 1.4 The clustering stage and keying process to protect from different attacks.

group indicated by DK_b. Finally, let $x_b^s = (x_{b1}, \ldots \ldots x_{bn})$ be the b-th row of S that contains the level of consistency of x_{bn} for each cluster. The level of consistency and the time taken for processing is shown below:

$$X_{a,b} \in (0,1), a = 1, \ldots \ldots M; b = 1, \ldots .. m \tag{1.5}$$

The level of consistency $X_{a,b}$ and the time taken for processing the encryption is obtained from Equation (1.5); here, a, b are represented as different groups of vectors. M, m represent the network configuration.

Nodes send a packet to their entire network to identify their neighbors and use it as a biometric-based asking communication to commence the keying step after deployment is complete. An encoded group and a sequence number are included in the packet delivered to every node, and the energy usage for each node and the cluster head is calculated. In the storage stage, nodes in the network store acquired requests, their respective identifiers, and the encoded information that goes with them. Employing biometric characteristics and dynamic variables, a network mx_a starts the construction of a physiological pairing key together with its companions. It is possible to keep the life duration of a volatile key to a minimum by employing this technique.

A biometrically verified telecast transmission prevents an offender from making a misleading bilateral demand. Otherwise, the node would wait until it receives a request from a neighboring node. Once la_u has been obtained by the node mx_a it can start a process of encrypting and decrypting data with either one of its friends, such as mx_b and then transmits to mx_a. In a simple style, the request includes its identification and selected group. As a result, a similar key is generated from every set of adjacent nodes based on the circumstances described above. The attacks in the sensor network have been protected by disclosing the shared key between mx_b and mx_a is shown below:

$$S_{ab} = S_b \text{ attacks } (mx_a \cup la_u \cap mx_b) \tag{1.6}$$

The protection stage of networks from different attacks is obtained from Equation (1.6); here mx_a, mx_b different keys in the nodes, la_u represent the request from the neighbor node. The use of a fuzzy membership criterion, in this case, is intended to increase safety. The membership value generates new prime numbers before performing the collision operation. A rectangular association value is used to generate new unique quantities. C-means clustering provides superior results compared to other algorithms, especially for overlapping data sets. In contrast to k-means, every other data point should correspond to exactly one cluster center; in this case, data points are awarded membership to all cluster centers. The BBUCAF increases the security level with the increase in the speed of the network with less network traffic and protection against various attacks.

1.4 Experimental Analysis

Modeling a network with many sensors is used to evaluate the system's privacy and efficiency. To measure the performance of various methods, a few sensor networks to overcome the attacks are used with high-speed performance with less traffic. The network's performance is increased using less traffic among the networks and protection against attacks; the evaluation of the network in terms of traffic is measured in the form of latency. The performance of traffic among the network is shown in Figure 1.5.

Computational calculations show that the network's activity is implemented in this simulation scenario. For example, parameters such as the percentage of assaults detected and the time it takes to respond to well-known sensor network attacks can be calculated.

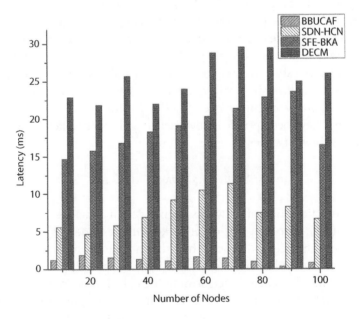

Figure 1.5 Latency of BBUCAF.

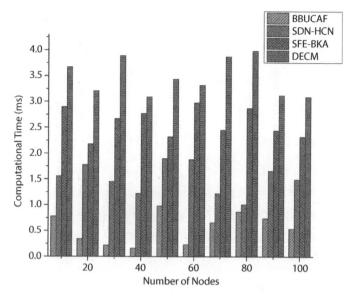

Figure 1.6 Computational time of BBUCAF.

Figure 1.6 shows the outcomes of an analysis of various strategies concerning the operating duration of the system under different node counts. The suggested BBUCAF system takes less time to execute than alternative approaches. Biometric key authentication is used to secure more communication, resulting in a running time of a few milliseconds for different nodes.

Eu is used to calculating the total energy usage of the network as shown below.

$$Eu = (l + 1) * m(12) \qquad (1.7)$$

The total energy consumption is obtained from Equation (1.7), shown in Figure 1.7; here, l represents the sensor networks that help send m messages between nodes. As a result, the efficiency of energy use is directly related to the amount of time spent concentrating, the size of the packet sent, and the amount of time spent decrypting and encrypting data.

With the delay, one may determine the typical end-to-end lag experienced by data packets as they travel across networks. The term "end-to-end delay" refers to the average amount of time it takes for a packet delivered from a resource to reach its target. Trails are used to calculate delay, as shown in Figure 1.8.

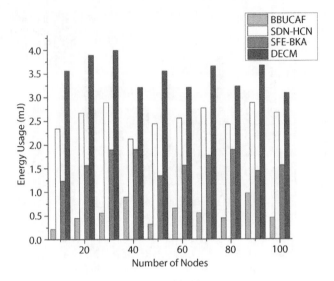

Figure 1.7 Energy usage of BBUCAF.

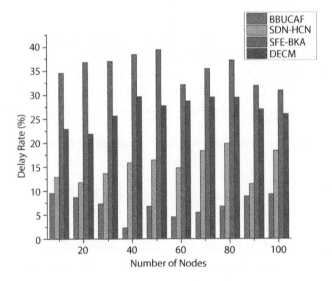

Figure 1.8 Delay rate of BBUCAF.

$$delay = \Sigma_l \, Q_{de}/m_l \qquad (1.8)$$

The delay for the entire network is obtained from Equation (1.8); here l represents different sensor networks, m_l represent the total number of packets collected.

Figure 1.9 compares the outcomes of different strategies for attack detection based on the number of nodes. Compared to previous approaches, the attack detection rate of the proposed BBUCAF system is more advanced.

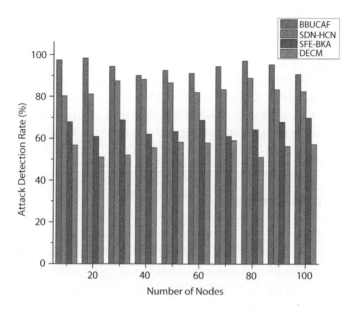

Figure 1.9 Attack detection rate of BBUCAF.

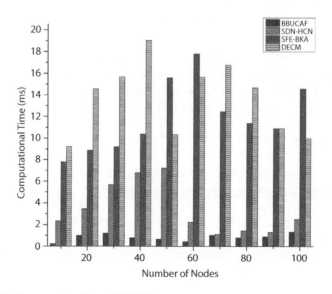

Figure 1.10 Computational time BBUCAF.

A biometric key authentication method is used to secure more interaction since the projected system has a lower attack detection performance than the present system.

To test the effectiveness of different methods, the following notations are used for the multiplication time with the hash function and completion time. The fuzzy extractor and the computation time relate to multiplying rates in the algebraic graph, and the hash function with the cost of a unique action may be disregarded bitwise. The BBUCAF uses the clustering node to decrease the disparity between the computational loads of the center and the nodes and can enhance the efficiency of the performance. The computational time of BBUCAF is shown in Figure 1.10. As a result, just the expenses of calculating irregular multiplier curve operations and a hash function should be considered for detecting the computational cost.

1.5 Conclusion

To deal with the security-related problems, Network security and speed have been improved with the introduction of a BBUCAF. The fingerprint is used to develop a Biometric-based model. The user's private key is generated securely using feature vectors. Sensor nodes get a key in the form of an encrypted message. The user's key is given to each sensor node, and a random number is used to create private keys between them. A fuzzy registration component that considers prime integers is used to create a collective key. Based on biometrics, data is sent between groups and sensor nodes using fuzzy membership and secret keys. The set of group keys used by each cluster is distinct from the sets used by other clusters. To increase the network's effectiveness, the network's speed must be increased to reduce network traffic with less energy usage and guard against DoS attacks.

References

1. Dowlatshahi MB, Rafsanjani MK, Gupta BB. An energy-aware grouping memetic algorithm to schedule the sensing activity in WSNs-based IoT for smart cities. *Applied Soft Computing.* 2021 Sep 1;108:107473.
2. Kumar A, Sharma K, Singh H, Naugriya SG, Gill SS, Buyya R. A drone-based networked system and methods for combating coronavirus disease (COVID-19) pandemic. *Future Generation Computer Systems.* 2021 Feb 1;115:1-9.

3. Dhasarathan C, Kumar M, Srivastava AK, Al-Turjman F, Shankar A, Kumar M. A bio-inspired privacy-preserving framework for healthcare systems. *Journal of Supercomputing*. 2021 Oct;77(10):11099-134.
4. Khan A, Gupta S, Gupta SK. Emerging UAV technology for disaster detection, mitigation, response, and preparedness. *Journal of Field Robotics*, 2022 Sep;39(6):905-55.
5. Je SM, Huh JH. Estimation of future power consumption level in smart grid: Application of fuzzy logic and genetic algorithm on the big data platform. *International Journal of Communication Systems*. 2021 Jan 25;34(2):e4056.
6. Nayak P, Swetha GK, Gupta S, Madhavi K. Routing in wireless sensor networks using machine learning techniques: Challenges and opportunities. *Measurement*. 2021 Jun 1;178:108974.
7. Kumar V, Malik N, Dhiman G, Lohani TK. Scalable and storage efficient dynamic key management scheme for wireless sensor network. *Wireless Communications and Mobile Computing*. 2021 Jul 1;2021.
8. Rakhra M, Bhargava A, Bhargava D, Singh R, Bhanot A, Rahmani AW. Implementing machine learning for supply-demand shifts and price impacts in farmer market for tool and equipment sharing. *Journal of Food Quality*. 2022 Mar 15;2022.
9. Karunanithy K, Velusamy B. Directional antenna based node localization and reliable data collection mechanism using the local sink for wireless sensor networks. *Journal of Industrial Information Integration*. 2021 Dec 1;24:100222.
10. Yu S, Gong X, Shi Q, Wang X, Chen X. EC-SAGINs: Edge computing-enhanced space-air-ground integrated networks for internet vehicles. *IEEE Internet of Things Journal*. 2021 Jan 19.
11. Karvelis P, Georgoulas G, Kappatos V, Stylios C. Deep machine learning for structural health monitoring on ship hulls using acoustic emission. *Ships and Offshore Structures*. 2021 Apr 21;16(4):440-8.
12. Benedetti P, Femminella M, Reali G, Steenhaut K. Experimental analysis of the application of serverless computing to IoT platforms. *Sensors*. 2021 Jan;21(3):928.
13. Wei D, Ning H, Shi F, Wan Y, Xu J, Yang S, Zhu L. Dataflow management in the internet of things: sensing, control, and security. *Tsinghua Science and Technology*. 2021 Jun 9;26(6):918-30.
14. Chen Y, Chen J. An efficient mutual authentication and key agreement scheme without password for wireless sensor networks. *Journal of Supercomputing*. 2021 Dec;77(12):13653-75.
15. Kui X, Feng J, Zhou X, Du H, Deng X, Zhong P, Ma X. Securing top-k query processing in two-tiered sensor networks. *Connection Science*. 2021 Jan 2;33(1):62-80.
16. Baumann H, Lindkvist M. A sociomaterial conceptualization of flows in industrial ecology. *Journal of Industrial Ecology*. 2022 Apr;26(2):655-66.

17. Wu Y, Ma Y, Dai HN, Wang H. Deep learning for privacy preservation in autonomous moving platforms enhanced 5G heterogeneous networks. *Computer Networks*. 2021 Feb 11;185:107743.
18. Wang C, Wang D, Xu G, He D. Efficient privacy-preserving user authentication scheme with forward secrecy for industry 4.0. *Science China Information Sciences*. 2022 Jan;65(1):1-5.
19. Garba A, Dwivedi AD, Kamal M, Srivastava G, Tariq M, Hasan MA, Chen Z. A digital rights management system based on a scalable blockchain. *Peer-to-Peer Networking and Applications*. 2021 Sep;14(5):2665-80.
20. Khan T, Singh K, Hasan MH, Ahmad K, Reddy GT, Mohan S, Ahmadian A. ETERS: A comprehensive energy aware trust-based efficient routing scheme for adversarial WSNs. *Future Generation Computer Systems*. 2021 Dec 1;125:921-43.
21. Haseeb K, Din IU, Almogren A, Ahmed I, Guizani M. Intelligent and secure edge-enabled computing model for sustainable cities using green internet of things. *Sustainable Cities and Society*. 2021 May 1;68:102779.
22. Zanelli F, Castelli-Dezza F, Tarsitano D, Mauri M, Bacci ML, Diana G. Design and field validation of a low power wireless sensor node for structural health monitoring. *Sensors*. 2021 Jan;21(4):1050.
23. Rubes O, Chalupa J, Ksica F, Hadas Z. Development and experimental validation of self-powered wireless vibration sensor node using vibration energy harvester. *Mechanical Systems and Signal Processing*. 2021 Nov 1;160:107890.
24. Wu TY, Yang L, Lee Z, Chu SC, Kumari S, Kumar S. A secure three-factor Authentication protocol for wireless sensor networks. *Wireless Communications and Mobile Computing*. 2021 Apr 16;2021.
25. Devi PP, Jaison B. Protection on wireless sensor network from clone attack using the SDN-enabled hybrid clone node detection mechanisms. *Computer Communications*. 2020 Feb 15;152:316-22.
26. Mahendran RK, Velusamy P. A secure fuzzy extractor based biometric key authentication scheme for body sensor network in Internet of Medical Things. *Computer Communications*. 2020 Mar 1;153:545-52.
27. Ashraf S, Ahmed T, Aslam Z, Muhammad D, Yahya A, Shuaeeb M. Depuration based Efficient Coverage Mechanism for Wireless Sensor Network. *Journal of Electrical and Computer Engineering Innovations (JECEI)*. 2020 Jul 1;8(2):145-60.
28. Wu F, Li X, Xu L, Vijayakumar P, Kumar N. A novel three-factor authentication protocol for wireless sensor networks with IoT notion. *IEEE Systems Journal*. 2020 Apr 28;15(1):1120-9.

29. Rangwani D, Sadhukhan D, Ray S, Khan MK, Dasgupta M. An improved privacy-preserving remote user authentication scheme for agricultural wireless sensor network. *Transactions on Emerging Telecommunications Technologies*. 2021 Mar;32(3):e4218.
30. Yu S, Park Y. SLUA-WSN: Secure and lightweight three-factor-based user authentication protocol for wireless sensor networks. *Sensors*. 2020 Jan;20(15):4143.

2

DeepNet: Dynamic Detection of Malwares Using Deep Learning Techniques

Nivaashini, M.[1]*, Soundariya, R. S.[2], Vishnupriya, B.[3] and Tharsanee, R. M.[4]

[1]Department of Computer Science & Engineering, Sri Ramakrishna Engineering College, Coimbatore, Tamil Nadu, India
[2]Department of Computer Technology, Bannari Amman Institute of Technology Sathyamangalam, Tamil Nadu, India
[3]Department of Computer Science & Engineering, KPR Institute of Engineering and Technology, Coimbatore, Tamil Nadu, India
[4]Department of Computer Science & Engineering, Bannari Amman Institute of Technology, Sathyamangalam, Tamil Nadu, India

Abstract

The innovation of technologies has become ubiquitous and imperative in day-to-day lives. Consequently, there has been a massive upsurge in malware evolution, which generates a substantial security hazard to organizations and individuals. This advancement in the competencies of malware opens new cybersecurity research dimensions in malware detection. It is quite impossible for anti-virus applications using traditional signature-based methods to find novel malware that incurs high overhead with respect to memory and time. This is because malware developers explore new methodologies to avoid these traditional malware defense approaches. To solve the problem, machine learning algorithms are used to learn the distinctions between malware and benign apps automatically. Unfortunately, traditional machine learning approaches that are constructed on handmade features are rather ineffective against these elusive practices and need more efforts owing to feature-engineering. To overcome such limitations, this work proposes a well-defined malware detection system called DeepNet based on deep learning techniques. In this work, we focus on the application of deep learning frameworks for malware detection by evaluating their effectiveness when malware is represented by high-level and low-level features, respectively. In this paper, two deep learning models, Stacked Autoencoder (SAE) and Deep Belief Networks (DBN)

Corresponding author: nive19794@gmail.com

S. Sountharrajan, R. Maheswar, Geetanjali Rathee, and M. Akila (eds.) *Wireless Communication for Cybersecurity*, (21–54) © 2023 Scrivener Publishing LLC

with Restricted Boltzmann Machine (RBM) are utilized to extract better features. SoftMax and Deep Neural Networks (DNN) classifiers are utilized in the malware discovery and classification. Comprehensive experiments are achieved on four benchmark malware datasets namely, Malimg dataset, BIG 2015 dataset, MaleVis dataset, and Malicia dataset. The outcomes implies that the proposed hybrid architecture can sense new malware trials with improved correctness and minimal false positive rates in comparison with the conservative malware prediction systems while preserving least computational time. The proposed hybrid framework is also unfailing and operative against complication outbreaks in malware recognition.

Keywords: Malware detection, deep learning, attribute reduction, feature engineering, dimensionality reduction

2.1 Introduction

Modern computer technology and the Internet have made life simpler and easier for people. Nowadays, anything can be done online, including social contact, financial transactions, tracking changes in the human body, etc. These kinds of advancements tempt cybercriminals to commit crimes online rather than in the actual world [1]. Recent scientific and commercial publications estimate that cyberattacks cost the global economy trillions of dollars. Malware is a common tool used by online criminals to start attacks. Any software known as malware engages in unauthorized and suspicious actions on the computers of its victims. The different varieties of malware include viruses, worms, Trojan horses, and ransomware which can steal sensitive information, launch assaults, and cause havoc to computer systems [2]. The latest malware iterations hide themselves on the victim's system by encrypting data and stuffing it. These novel varieties propagate by using people's trust as a vehicle for infection. For instance, well-known malware transmission vectors are launched through email attachments, viewing, and downloading files from bogus websites. To keep computer systems safe, we must identify malicious software as soon as it affects the systems. Malware classification is the process of examining and locating files to determine whether they are malicious or non-harmful [3].

Machine learning technologies along with cloud computing and block chain are all used in these procedures to boost the detection rate. Using the methods and tools, there are various malware detection strategies. The key methods used here include memory-based and model detection, model validation, behavior, and signature analysis [4]. Depending on the methods and tools employed, various approaches have different names. The use of a signature-based strategy works well against known and related malware variants. However, it is unable to find malware that has not yet been seen. Although the other detection systems can successfully identify some unknown malware components, they fall short when it comes to identifying sophisticated malware variants that employ packaging and concealment tactics [5].

The inadequacies of current malware recognition technologies have been solved in recent times using a deep learning–based approach. Numerous fields, including image processing, NLP, human action, and facial recognition have made substantial use of deep learning [6]. However, deep learning has got more roles in the field of cybersecurity, particularly in malware recognition. Artificial neural networks are the foundation of the subset of artificial intelligence known as deep learning. Deep learning learns from past instances and employs numerous hidden layers. A variety of deep learning architectures, including deep neural networks (DNN), recurrent neural networks (RNN), convolutional neural networks (CNN) and deep belief networks (DBN), have been employed to improve model performance [7].

In this work, a unique hybrid deep learning approach, DeepNet, has been proposed for classifying malware. The proposed model is trained using three different datasets such as Malimg, MaleVis and BIG 2015. The input malware images in these datasets are initially converted as binaries before it is fed to the training process. After conversion, the bit/byte level sequences are forwarded to the feature extraction process. Stacked Auto Encoders (SAE) along with Deep Belief Networks (DBN) and Restricted Boltzmann Machine (RBM) are used for extracting the features. The malware classification is implemented using the softmax classifier. According to the test results, the suggested method may successfully extract distinguishing characteristics for each type and family of malware in order to classify it. The findings of the experiment also demonstrated that the suggested deep learning algorithm classifies several malware variants with excellent accuracy, outperforming the most recent methods described in the literature.

The novelty of the proposed work is as per the following:

i) A powerful and quick DL-based malware acknowledgment framework utilizing crude parallels while requiring no paired execution (conduct investigation), picking apart, or code dismantling language abilities is given.

ii) The proposed crossover model utilizes pretrained Profoundly Associated DBN with RBM and SAE (DeepNet) to accomplish quicker preprocessing and preparing of parallel examples. The DeepNet model licenses for combination of highlights and uses less boundaries contrasted with other DL models. The selective profound management component of the DeepNet model gives to successful malware disclosure. Moreover, the profound associations with its normalizing power help decrease overfitting with diminished malware preparing tests.

iii) The issue of information lopsidedness in sorting malware is attempted by reweighting of the class-adjusted unmitigated cross-entropy misfortune capability in the softmax layer.
iv) We direct a broad assessment on four different malware datasets, of which three datasets are utilized for preparing and one dataset is utilized for testing the proposed model. The results show that the proposed structure is extremely strong and capable. It is additionally vigorous against modern malware improvement over the long run and in consistency to hostile to malware avoidance strategies.
v) The proposed mixture system accomplishes higher precision paces of 98.7%, 98.5%, and 98.2% for the three datasets and of 90.2% for the concealed (Malicia) dataset. The model achieves expanded computational execution with diminished reality intricacy, in this way achieving a useful malware acknowledgment framework.

The rest of this paper is coordinated as follows. Segment 2.2 portrays the malware distinguishing proof and grouping techniques examined in the writing. Segment 2.3 spotlights the datasets that are accessible for malware location. Segment 2.4 shows the profound structures that are reasonable for malware recognition. Segment 2.5 depicts the proposed DeepNet model with engineering. Segment 2.6 presents the trial consequences of the DeepNet model and looks at the outcomes acquired against other AI and DL models. Segment 2.7 concludes the current work.

2.2 Literature Survey

Malware examination can be arranged into static and dynamic sorts in two principal bunches [8]. Both manual and mechanized investigation is conceivable. While programmed investigation requests huge information science programming capacities, manual examination requires subject aptitude. Static examination of malware is the initial step followed by unique investigation, which comes last. The static investigation recognizes the design of the malware test without running the genuine noxious projects. The reason and usefulness of malware are uncovered through cutting edge static investigation, which calls for profoundly specific information on working framework ideas and get together code directions.

Dynamic examination includes running projects and dissecting malware's exercises. Contrasted with static examination, dynamic investigation more definitively portrays the genuine capacities of malware. The two classes of dynamic examination are essential powerful investigation and high-level unique investigation. Essential unique examination utilizes observing instruments while cutting-edge dynamic investigation utilizes troubleshooting apparatuses to look at the exercises of the malware.

The crossover examination came into execution to defeat the disadvantages of static and dynamic investigation. Static examination will check for application and client authorization, and wary code though powerful investigation checks for the way of behaving of the application. In this strategy, to improve malware examination, dissecting any malevolent code's mark and joining it with other standard of conduct factors.

Without the user's knowledge, a computer hacker will send malware, open a loophole, and start processing bitcoin, a source of cryptocurrency, on the user's system. Malware that resides on the hard drive and runs in memory is either not verified, or there is a strong possibility that the malware's signature and behavioral pattern [9] differ from the malware that resides on the hard drive and runs in memory.

2.2.1 ML or Metaheuristic Methods for Malware Detection

High-performance RF, KNN, and AdaBoost with prior research indicating successful RF and KNN mobile malware detection. In the signature-based malware detection method, in which signatures are extracted and compared, and based on the comparison they are classified as malicious. In automatic malware detection [10], the string signatures were automatically retrieved using a variety of library recognition methods and diversity-based criteria. The application contains various cryptographic hash-based signatures in accordance with the tamper-evident architecture. These signatures allow for the detection of Trojans hidden within the hardware.

There are certain drawbacks of Machine learning algorithms used in detecting malwares [11].

1. Adaboost has good new discovery detection capabilities and gave malware version 8 a perfect score.

2. Due to the increased noise in our dataset, using real-world data for model training may have resulted in worse performance.
3. When using Machine Learning algorithms like KNN, Bayes Network and so on, the False-negative rate (malicious detected as malicious) is lower than the false-positive rate (benign wrongly detected as malicious).
4. If the size of the dataset is smaller then AdaBoost will give a good accuracy rate in finding Ransomware.
5. For tracking real-time, malicious, or suspicious data which requires a long time when the KNN algorithm is used, as smartphones will have less computation power; hence KNN is not applicable from mobile phones.
6. The inadequate training instances may consequently lower the performance on the identification of Spyware SMS and Adware.
7. F1 scores that are lower than those of other cutting-edge dynamic malware detectors.
8. False Negative rate is higher, which means most of the suspicious actions are not detected.

Most ML calculations give great precision, yet that relies upon the dataset and how the model is prepared to find the malware. Metaheuristic calculations are utilized for distinguishing malware that have profoundly connected highlights, not every one of the elements an enhanced methodology. Yet, this ML or Nature-propelled calculation isn't effective for continuous dataset or for cell phones; consequently, profound advancing should be carried out for taking care of a significant number of the dataset and the framework with less calculation power.

2.2.2 Deep Learning Algorithms for Malware Detection

A variant of machine learning called "deep learning" learns the input at many levels to provide improved knowledge representations. Convolutional Neural Networks (CNN) have been developed to advance computer vision through deep learning. Deep learning models train a complicated model with numerous convolutional layers and millions of parameters by learning complex features.

Regarding the amount of time and machine configuration needed for the experiment, LSTM [12] was the right approach. The LSTM layer is successively given the hexadecimal samples of clean wares and malware that have been transformed to numerical values by the input layer. Our model employs a stateful LSTM, which aids in identifying relationships among the text sequences that were taken from the dataset files' hex dump. About 128 neurons make up the LSTM layer, which oversees taking in inputs from the layer above and producing outputs using a linear activation function. Following multiple experiments, the network's features, time steps, and the number of neurons were decided upon.

Like an LSTM, Gated Recurrent Unit (GRU) [13], a potent variation of a conventional Recurrent Neural Network (RNN), uses the integrated gating mechanism as a short-term memory solution. A system inside the GRU called gates controls and even circulates the information flow. The gates assist the GRU cell in learning which information is crucial to store or remove. As a result, crucial information is passed on to enable prediction.

Convolutional neural networks play a major role in cyber security. Compared to traditional feature selection algorithms, CNN [14] can automatically learn the crucial features. CNN is viewed as a series of connected processing elements created with the goal of converting a set of inputs into a set of desired outputs. Convolution, pooling, flattening, and padding are just a few of the operations that CNN runs on the input data before connecting to a fully connected neural network. The CNN architecture's performance depends on its capacity to recognize and combine local input patterns in a parameter-effective manner. The CNN analyses an app's opcodes as text to be mined for malware-indicating patterns, concentrating on extracting n-gram characteristics from these sequences.

Conduct-based DL structure comprise Stacked Auto Encoders (SAE) [15] one of the most upgraded profound gaining calculations for malware discovery that takes criticism from conduct diagrams. The cloud stage (CP) and internet of things (IoT) climate (IoTE) modules assume an essential part in BDLF. The far-off PCs and other smart gadgets that make up the IoTE module communicate examining information or dubious recently introduced records to the focal handling unit (CP) and find solutions from the CP. The errand of recognizing examining information or records sent from IoTE falls on the locators in CP. For checking information, CP assembles conduct diagrams, changes over the Programming interface call charts into parallel vectors, and afterward takes care of the twofold vectors into SAEs models for malware recognizable proof. CP runs tests in the Cuckoo Sandbox and afterward pulls Programming interface calls from the checking documents of the sandbox for any dubious records. From that point forward, CP handles the observing information similarly to how it handles the filtering information. Following the location, CP illuminates IoTE. Programming interface call diagrams are made with the goal that they can consolidate Programming interface solicitations to learn perilous way of behaving.

A quick DL-based malware acknowledgment technique utilizing crude twofold snaps is presented by Thickly Associated Convolutional Organizations (DenseNet) [16], which includes no information on figuring out, parallel execution, or code dismantling. In contrast with past CNN models, the DenseNet model purposes less boundaries and grants the connection of highlights. Improved malware recognition is worked with by the DenseNet model's implicit profound oversight system. Furthermore, with decreased malware preparing datasets, the thick associations' regularizing limit lessens overfitting.

2.3 Malware Datasets

2.3.1 Android Malware Dataset

CICMalDroid 2020 [17], the data consist of 17,341 samples (from 2017 to 2018), the source includes the Contagio security blog, Virus Total service, MalDoz, AMD, and from various dataset used for cyber research. It is important for cybersecurity experts to categorise Android apps as malware in order to implement effective mitigation and countermeasure procedures. Therefore, we purposefully divided our dataset into five different categories: Banking malware, SMS malware, Adware, Riskware, and Benign.

2.3.2 SOREL-20M Dataset

SOREL-20M [18] makes an overall or partial attempt to address these problems. By offering orders of magnitude more data for analysis, we address the problem of training size. Internally, we have discovered that, although performance becomes better with bigger datasets, establishing a stable rank order amongst models and evaluating performance with fewer false positives only requires validation sizes of about 3 to 4 million cases. We get 12,699,013 training samples, 2,495,822 validation samples, and 4,195,042 test samples when our suggested time divides are applied to create the training, validation, and test sets, respectively. LightGBM and a PyTorch-based feed-forward neural network (FFNN) model using SOREL-20M. Although both models perform well, there is still much opportunity for improvement, especially at lower false positive rates. As a result, SOREL-20M should be more valuable to contrast various malware detection strategies. In addition, we present benchmarks for employing a multi-target model and a variety of extra targets that define behaviors inferred from vendor labels.

2.4 Deep Learning Architecture

The basic deep learning architectures suitable for malware detection are described briefly in this section.

2.4.1 Deep Neural Networks (DNN)

A conventional artificial neural network with numerous interconnected layers between the input and output layers is known as a DNN [19]. In order to transform the input into the output, the DNN determines the appropriate mathematical computation. Numerous neurons in the single layer of the DNN are where computations are done. The node accepts input, processes it using stored weights, applies an activation function, and then passes the results to the next node in line until a conclusion is reached as shown in Figure 2.1.

2.4.2 Convolutional Neural Networks (CNN)

CNN is presently thriving in the realm of cyber security after attaining remarkable results in the disciplines of image recognition, audio recognition and computer vision [25]. Compared to traditional feature selection algorithms, CNN can automatically learn the crucial features [9]. CNN

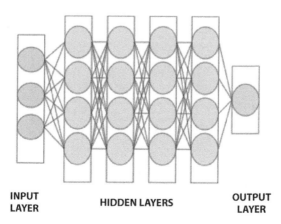

Figure 2.1 Architecture of DNN.

is viewed as a series of linked processing elements designed to convert a set of inputs into a set of desired outputs. The three primary parts of a CNN classifier are input, output, and hidden layers as shown in Figure 2.2. Convolution, pooling, flattening, and padding are the fundamental operations that CNN runs on the input data before connecting to a fully connected neural network.

2.4.3 Recurrent Neural Networks (RNN)

The RNN variation known as Long Short-Term Memory (LSTM) is a remarkable classifier to learn and mine temporal data. In order to learn long-term characteristics and relationships, the LSTM model makes use of an exceptional module [10]. Additionally, using different "gate" structures reduces reverse propagation of error. The "gate" state, which controls the data stream and memory, determines the internal values of the exceptional module based on the information currently available and prior flows. There are three gates present in each LSTM cell such as Input gate, forget gate, output gate. Additionally, two different states are also available which are hidden states and the cell states as shown in Figure 2.3.

2.4.4 Deep Belief Networks (DBN)

A DBN is a Restricted Boltzmann Machine (RBM) that is layered as a self-organizing graphical network. RBM is an undirected comprehensive model in which the modules of the same layer are not connected; only

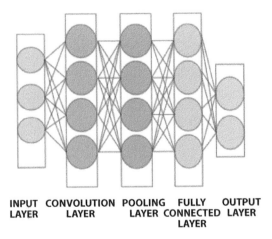

Figure 2.2 Architecture of CNN.

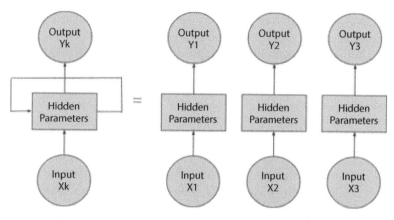

Figure 2.3 Architecture of RNN.

the layers are related [35]. The construction of a DBN is very straightforward; RBMs are layered to create an unsupervised network that treats the visible and hidden layers as separate networks. The following layer, and so on, considers this hidden layer to be a visible layer. The met discourse delineation is stratified for each RBM in each sub network, and the layers act as feature extractors. After pretraining, supervised learning is used for refining, which enables the DBN to perform binary classification in order to determine whether the input features correspond to malicious software or a benign application.

2.4.5 Stacked Autoencoders (SAE)

A heap of autoencoders acting as hidden layers in a neural network is known as a stacked autoencoder. A back propagation mechanism is utilized by the stacked autoencoder, an unsupervised machine learning method, to project the output value [36]. With noisy autoencoders built into the layers, this neural network enhances accuracy in deep learning. The following three steps are typically included in stacked autoencoders.

Stage 1: The information is utilized to prepare the autoencoder, which then creates the obtained highlights.
Stage 2: The accompanying layer involves these elements as an info, etc., until the preparation is done.
Stage 3: Assuming that the secret layer is prepared, the back-engendering calculation (BP) is utilized to bring down the expense capability and revamp the loads by marking preparing information to accomplish wanted execution.

The proposed work integrates Profound Conviction Organizations with RBM and SAE as a cross-breed model to separate wise elements from the dataset before sending it for the malware grouping utilizing softmax classifiers. The proposed framework utilizes RBM with DBN and SAE over the other DL calculations for the following reasons:

i) The proposed DeepNet model utilizes three datasets with various aspects, that thus might influence the functioning effectiveness of the proposed framework. In this spot RBM assumes the significant part of dimensionality decrease for diminishing the quantity of irregular factors to a bunch of standard factors in the malware datasets.

ii) Being a generative model permits DBNs to be utilized in either an unaided or a regulated setting. Meaning, DBNs in the proposed framework are likewise utilized for highlight learning/extraction. Exactly, in highlight learning we do layer-by-layer pre-preparing in a solo way on the different RBMs that structure a DBN.

iii) Each RBM model plays out a non-straight change on input vectors and produces as results vectors will act as contribution for the following RBM model in the sequence. This permits a ton adaptability to DBNs and makes them simpler to extend.

iv) The SAE is a nonlinear change to find the primary component heading, during the time spent include learning/extraction and DBN depends on the likelihood of dispersion of tests to remove significant level portrayals.

v) The fundamental components of both scanty autoencoder and RBM are different on a basic level. In the preparation strategy, the SAE for the most part involves the angle plunge technique same as DBN with RBM. The general progression of the SAE and DBN preparing is predictable, with a layer of preparing.

2.5 Proposed System

2.5.1 Datasets Used

The proposed system considers three datasets, namely Malimg [23], BIG 2015 [24] and MaleVis [22] datasets. The Malimg dataset contains images of

Table 2.1 Datasets used in the proposed system.

Dataset	Family/Classes	Total samples	Training samples	Testing samples
Mailmg [20]	Adialer.C ,Agent.FYI , Allaple.A , Allaple.L, Alueron.gen!J, Autorun.K , C2LOP.gen!g, C2LOP.P, Dialplatform.B , Dontovo.A, Fakerean, Instantaccess, Lloyds.AA1, Lloyds.AA2, Lloyds.AA , Lolida.AT, Malex.gen!J, Obfuscator.AD, Rbot!gen , Skintrim.N, Swizzor.gen!E, Swizzor.gen!I, VB.AT, Wintrim.BX, Yuner.A	9339	6437	2115
BIG 2015 [21]	Ramnit, Lollipop, Kelihos_ver3, Vundo, Simda, Tracur, Kelihos_ver1, Obfuscator. ACY, Gatak	21741	8338	3573
MaleVis [22]	Vilsel, VBKrypt, VBA/Helium.A, Stantinko, Snarasite.D!tr, Sality, Regrun.A, Neshta, Neoreklami, MultiPlug, InstallCore.C, Injector, Hlux!IK, HackKMS.A, Fasong, Expiro-H, Elex, Dinwod!rfn, BrowseFox, AutoRun-PU, Androm, Amonetize, Allaple.A, Agent-fyi, Adposhel, AutoRun-PU, Androm, Amonetize, Allaple.A, Agent-fyi, Adposhel	14226	9958	4268

9,339 malware, each of them belonging to 25 families; the dataset is helpful for malware classification in terms of multiple classes. The BIG 2015 dataset containing the malware binary samples of about 21,741 was introduced by Microsoft for malware classification, each of which represents 9 divergent families. The MaleVis malware dataset is an image dataset generated from 25 malware and one benign software classes, applicable for vision-based malware identification and it is specially designed for implementing deep learning architectures. The dataset comprises malware images of around 14,226 RGB, each of them belonging to the 26 classes. The malware images in the above datasets are converted into binaries, in order to avoid ambiguity among the input variables that are passed into the proposed DeepNet architecture for further processing. Table 2.1 Represents the training and testing samples for all the three datasets.

2.5.2 System Architecture

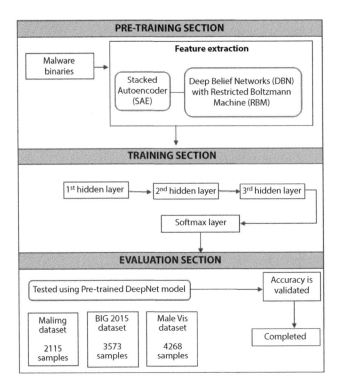

Figure 2.4 Overall design & flow of the proposed DeepNet model.

2.5.3 Data Preprocessing

Data preprocessing is the process of eliminating noise, missing values, and inconsistent data found in the dataset. The raw data is cleaned and then the data is transformed into a format suitable for further processing. The proposed system considers three datasets (refer to Table 2.1) that contains malware images, which are converted into malware binaries and then fed as input to the DeepNet algorithm. Initially the malware images are interpreted as 2D matrix, which is then converted into 8-bit vectors ranging from 0 to 255, followed by the conversion of vectors in binaries in the form of 1's and 0's. The proposed DeepNet model can handle RGB/gray scale images. But the main reason for binary conversion is that binaries simplify the algorithm and reduce computational requirements. However, processing RGB layers is more complex, Hence the converted malware binaries are used for further processing as shown in Figure 2.5. In a similar way, the training samples of all the three datasets are converted as binary files and are carried forward for the feature extraction process.

2.5.4 Proposed Methodology

The complete scheme of the anticipated malware recognition method is illustrated in Figure 2.4. The emergence of the anticipated amendment of DeepNet representation with DBN and SAE layers is shown in Figure 2.5 and algorithm 1. The key twofold images are nourished into the DeepNet representation for attribute mining and categorisation. The representation is accomplished through offering the twofold images precisely into the DBN and SAE layers. The anticipated DeepNet representation with DBN [34] and SAE [34] layers has a boundless ability to mine distinguishing attributes that broadly express the figure and study task-specific attributes.

They spontaneously study the attributes at numerous stages of extraction, permitting them to study dense purposes through demonstrating primal key information into the anticipated outcome. The anticipated prototype utilises DeepNet to mine the whole attributes from malware datasets then prepares the DeepNet on maximum of the mined attributes. Each deep level can mine good particulars from twofold images. The turnout attribute records obtained subsequently going around these levels are provided as a key for a fully connected (FC) level. The FC level categorises the malware trails into their related categories.

Algorithm 1. DeepNet algorithm
Input: Twofold image trials
Output: Precise class Ci
a. Convert binaries to 2D range grayscale imageries to the array I.
b. Prepare the prototype.
c. Mine untrained attributes from the key trials.
d. Arbitrary loading of key training trials.
e. Standardized training trials are provided to DeepNet network (DBN + SAE) and the number of the primary neural elements is established as the number of attributes in the key training trials.
f. Reiterate the procedure till the DeepNet is prepared to take on the requirements of repetition or the divergence state.
g. Connect every level through joining the attribute plots of entire former levels.
h. Categorise the key trials into their equivalent categories utilising a softmax classifier.

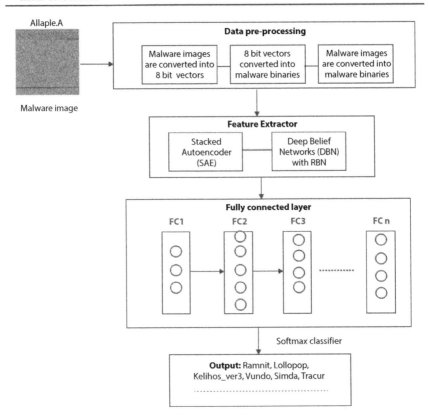

Figure 2.5 Architecture of the proposed DeepNet methodology.

2.5.5 DeepNet

DeepNet is a DL model in which whole levels are totally connected, in this way achieving a productive information stream among them. Each level acquires an additional inclusion from the entire going before levels then, at that point, moves its trait plots to the whole succeeding levels. The turn-out trait plots accomplished from the current level are combined with the first level through a chain. Each layer is associated with every one of the ensuing layers of the organization, and consequently authored as DeepNet. This model requires lesser cutoff points than traditional profound learning models. It additionally diminishes the overfitting downside that occurs with minor malware preparing gatherings.

For example, pondered on top, the information developments inside malicious applications could shift impressively from the friendly applications. DeepNet impacts such changes and similarities to suddenly perceive unique applications regardless of whether they are pernicious by using a DL model. Regular ML techniques, for example, Bayesian, SVM, MLP, and so on, ordinarily have less than three degrees of computation components. DL model, totally via its term shows, has a profound foundational layout including in excess of three hid layers. Its goals at producing a scholarly layered showing of the critical data to create useful properties for ordinary ML procedures. Each level in the level examinations an extra rundown and composite property of the data. General profound foundational layouts incorporate CNN, Scanty Coding, RBM, DBN, RNN, SAE, and so forth. In our examination, we picked DBN and SAE to build DeepNet.

2.5.6 DBN

DBN is a sort of DNN, comprising various degrees of covered factors called RBM, through joins among the levels among components inside each level. While zeroing in on a gathering of two-crease key picture preliminaries in an unaided means, the RBM levels in a DBN proceed as trait markers to concentrate on probability restoration of the property headings, which dynamically develop undeniable level portrayals. Hence, relating to the error among the key quality headings and the rebuilt bearings, meanings in the DBN are adjusted in a solo way. Later in the concentrating on stage, the DBN can be moreover achieved with sorted application preliminaries in a regulated means to execute categorisation. A back spread is precisely guided for calibrating to advance accuracy. In this mode, the DBN portrayal is completely built.

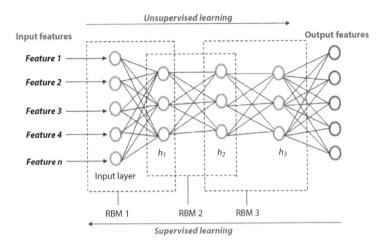

Figure 2.6 Structure of DBN model.

Our malware acknowledgment technique DeepNet is executed on the DBN engineering above, as displayed in Figure 2.6. Consequently, assembling and normalizing the property bearings as portrayed in the pre-handling section, DeepNet enters them into the DBN with RBM layers for dimensionality decreasing and characteristic mining. The DBN model is structured as stages shown in Figure 2.6. As this implies, DeepNet can ensure its accuracy in finding the ceaselessly developing new malware.

2.5.7 SAE

The framework of our expected plan applying SAE is introduced in Figure 2.7. The pre-handled credits are given as keys to the SAEs. There are 3 to 4 levels in the expected SAEs model. Relating to the plan of the framework level, the arrangement starts from level 1 autoencoder (AE) of the SAE model. We store the coordinating variables of level 1 AE to give a gainful essential impact for the readiness of level 2. When the level 2 frameworks are ready, the interesting data is at first placed to the level 1 framework to acquire the record of level 2. Correspondingly, the levels 3 and 4 are arranged unmistakably when the level 2 framework is ready.

When each level is arranged particularly in the framework, the streamlining activity assigned is Adam enhancer [33]. Towards advancing the exhibition of the model, a flowing strategy is used to improve the framework factors. Flowing the whole levels all in all cause a clever framework.

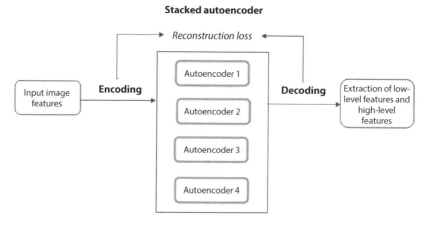

Figure 2.7 Structure of SAE model.

Right now, the yield of the underlying level is recovered. As the level 1 of the outpouring framework setup, the further levels are reimagined, then the elements like the impacts ready in each level are circulated. The proposed framework's exhibition affects the stacking levels of the AEs, which thusly corrupts in the event of stacking sequential degrees of AEs in the SAE model. Thus, considering the powerful contributions from the malware datasets the stacking levels might increment or decrement.

Later the whole model is ready, and we gain the finish of the SAE acknowledgment model. In the tertiary stage, we present our model tuning elements and fall in the getting ready technique. Our model is a brain network made out of AEs that get ready multi-facet frameworks level through level, that readies the complexity center of each and every level by means of AE toward the path from foremost to back. The yield of the last level is held in the job of key trait of the softmax classifier, and the categorisation results are yielded because of softmax.

2.5.8 Categorisation

The categorisation level is comprised of a completely associated (FC) SoftMax level. In FC, the number of neurons is fixed relating to the number of malware classifications introduced in the dataset. The SoftMax work is used for marking multi-class arrangement troubles. This works figures the probability divisions of each classification I upon the whole attainable classes.

The class disparity trouble is a categorisation struggle in which the division of classes in the getting ready dataset is unpredictable. The grade of classification disparity varies; a vital imbalance is further difficult to show and requires progressed techniques to deal with the issue. The Malimg dataset and the Microsoft Enormous 2015 dataset are ridiculous and broad - followed malware datasets that cover additional preliminaries for the minority classes and especially restricted preliminaries in specific classifications. Portrayals arranged on these blended preliminary aspects are impacted by principal sorts. To decide the issue of data disparity, data extension techniques like oversampling of more modest arranges or down inspecting of standard sorts are not reasonable for malware distinguishing proof inconveniences. It isn't likely to create portrayals identical to pragmatic malware doubles through oversampling. Various decisive malware options might be likely unseen by down inspecting.

Reweighting misfortunes through upset class event normally brings about down execution on useful data including a prevalent classification imbalance. The expected malware acknowledgment model uses class-adjusted misfortune [32] and uses a weighting component Wi, that has a reverse extent to the number of preliminaries for class I.

2.6 Result and Analysis

The dataset was for arbitrary reasons isolated into 70% preparation and 30% approval sets. The results were saved with 1,043 clean product preliminaries and every one of the three malware datasets. Train and test records were isolated to such an extent that 30% of the total preliminaries were read up for investigation conclusions. The expected malware acknowledgment structure was prepared on 6,437 preliminaries and tried on 2,115 preliminaries for the Malimg dataset with clean product preliminaries (9339 + 1043). By then, the model was prepared on 8.338 preliminaries and tried on 3.573 preliminaries from the Large 2015 dataset alongside clean product preliminaries (10,868 + 1043). On the MaleVis dataset, 9958 preliminaries were preparing preliminaries and 4268 were trying preliminaries.

The detailed trials were executed on a Linux framework with Intel® Xeon(R) central processor E3-1226 v3 at 3.30 GHz_4, 32 GB Slam, and NVIDIA GM107GL Quadro K2200/PCIe/SSE2. The execution valuations were taken out with the succeeding hyper parameter settings: 100 ages, learning rate 0.0001, and clump size 32. The expected profound brain network model was executed on the Python structure utilizing the TensorFlow Python library [26] and Keras v0.1.1 DL library.

There are four sorts of measurement systems estimated to evaluate class likelihoods.

  a. True Positive (TP): the likelihood that a sample fits to a group and it ensures fit to that group, i.e., a trial that is categorised as malware and is malware.
  b. True Negative (TN): the likelihood that a sample does not fit to a group and it does not fit to that group, i.e., a trail that is categorised as not malware and is not malware.
  c. False Positive (FP): the likelihood that a sample fits to a group and it does not fit to that group, i.e., a trial that is categorised as malware and is not malware.
  d. False Negative (FN): the likelihood that a sample does not fit to a group and it does fit to that group, i.e., a trial that is categorised as not malware and is malware.

Accuracy (Acc), Precision (Pr), Recall (Re), and F1 score are the four key categorisation systems of measurement. The number of precise likelihoods partitioned by the total number of likelihoods is known as accurateness. It is defined as

$$Acc = (TP+TN)/(TP+FP+TN+FN)$$

Precision is the number of precise definite results partitioned by the number of definite results anticipated by the classifier. It is defined as

$$Pr = TP/(TP+FP)$$

Recall provides the division of appropriately recognized occurrences as the definite outcomes of all the definite ones. It is given by

$$Re = TP/(TP+FN)$$

F1 score is the harmonic mean of precision and recall. It influences the classifier's precision along with its strength. It is given by

$$F1\ score = 2\times((precision \times recall)/(precision+recall))$$

The evaluation outcomes of conventional approaches for malware recognition are shown in Tables 2.2 and 2.3, Figures 2.8 and 2.9, respectively.

Table 2.2 Examination of ML-based techniques with the proposed DeepNet model for the three preparation datasets.

Models	Malimg dataset				BIG2015 dataset				MaleVis dataset			
	Acc (%)	Pr	Re	F-score	Acc (%)	Pr	Re	F-score	Acc (%)	Pr	Re	F-score
KNN	82.4	0.81	0.82	0.82	85.3	0.86	0.85	0.85	84.4	0.85	0.84	0.84
LR	69.2	0.70	0.67	0.68	62.6	0.64	0.62	0.63	66.5	0.67	0.66	0.66
SVM	75.1	0.75	0.75	0.75	89.3	0.90	0.88	0.89	88.4	0.88	0.87	0.88
NB	56.2	0.57	0.55	0.56	52.1	0.52	0.52	0.52	55.6	0.56	0.55	0.55
DT	88.4	0.88	0.87	0.88	86.4	0.86	0.86	0.86	87.4	0.88	0.87	0.87
RF	90.7	0.91	0.90	0.90	91.2	0.92	0.91	0.91	90.3	0.90	0.91	0.90
Adaboost	74.3	0.75	0.73	0.73	83.7	0.85	0.82	0.84	76.4	0.76	0.77	0.76
DeepNet	98.7	0.98	0.98	0.98	98.5	0.99	0.98	0.98	98.2	0.99	0.98	0.98

Table 2.3 Examination of DL-based techniques with the proposed DeepNet model for the three preparation datasets.

Models	Malimg dataset				BIG2015 dataset				MaleVis dataset			
	Acc (%)	Pr	Re	F-score	Acc (%)	Pr	Re	F-score	Acc (%)	Pr	Re	F-score
CNN	97.6	0.98	0.97	0.98	95.7	0.96	0.96	0.96	94.4	0.94	0.94	0.94
VGG16	97.4	0.98	0.97	0.97	88.6	0.89	0.89	0.89	96.2	0.97	0.96	0.96
VGG19	97.5	0.98	0.98	0.98	88.8	0.89	0.89	0.89	96.3	0.96	0.96	0.96
Inception-v3	97.7	0.99	0.99	0.99	93.3	0.93	0.93	0.93	95.3	0.96	0.95	0.95
Resnet-50	97.7	0.98	0.98	0.98	88.5	0.89	0.89	0.89	90.4	0.91	0.90	0.90
Xception	98.0	0.98	0.98	0.98	96.8	0.97	0.97	0.97	97.5	0.98	0.97	0.97
DenseNet-121	98.2	0.98	0.98	0.98	96.8	0.97	0.97	0.97	95.3	0.95	0.95	0.95
DeepNet	**98.7**	**0.98**	**0.98**	**0.98**	**98.5**	**0.99**	**0.98**	**0.98**	**98.2**	**0.99**	**0.98**	**0.98**

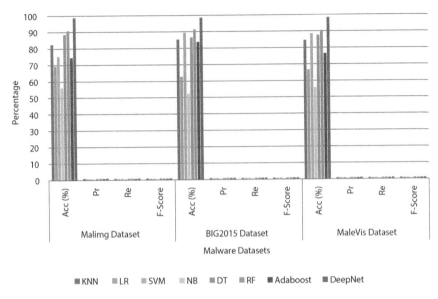

Figure 2.8 Comparison of ML-based techniques with the proposed DeepNet model.

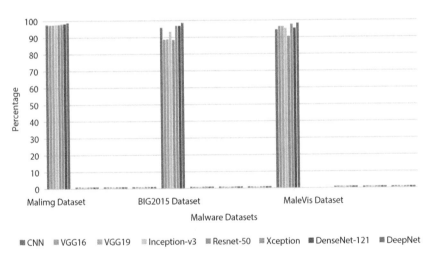

Figure 2.9 Comparison of DL-based techniques with the proposed DeepNet model.

The implementation investigation of the anticipated prototype is assessed with numerous ML methods like K-Nearest Neighbor (KNN), Logistic Regression (LR), Naïve Bayes (NB), SVM, Decision Tree (DT), Random Forest (RF), and Adaboost. The malware-based pretrained DL paradigms like CNN and its alternatives are utilised for examining the effectiveness of the anticipated DeepNet-based malware recognition technique.

Table 2.4 Examination of ML and DL models with the proposed DeepNet model for the Malicia (inconspicuous) dataset.

Methods		Acc (%)	Pr	Re	F-score
ML Methods	KNN	76.8	0.78	0.76	0.77
	LR	56.3	0.58	0.56	0.57
	SVM	80.3	0.81	0.80	0.80
	NB	50.0	0.46	0.47	0.47
	DT	77.8	0.78	0.77	0.77
	RF	82.1	0.83	0.82	0.82
	Adaboost	75.3	0.77	0.74	0.75
DL Methods	CNN	71.4	0.72	0.71	0.71
	VGG16	77.7	0.79	0.78	0.78
	VGG19	83.0	0.83	0.83	0.82
	Inception-v3	84.0	0.84	0.83	0.83
	Resnet-50	83.0	0.83	0.81	0.82
	Xception	83.0	0.83	0.81	0.82
	DenseNet-121	83.0	0.83	0.81	0.82
DeepNet		**90.2**	**0.90**	**0.90**	**0.90**

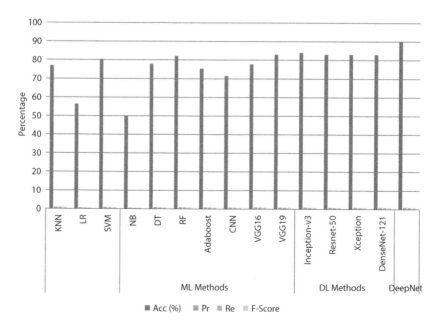

Figure 2.10 Comparison of ML & DL-based techniques with the proposed DeepNet model.

The implementation outcomes attained for the anticipated prototype are improved than the other traditional malware recognition approached for the three training datasets. The anticipated prototype attained an accurateness of 98.7% for Malimg, of 98.5% for BIG 2015, and of 98.2% for MaleVis dataset.

The over-simplification of the expected procedure is assessed through undetected dataset. The dataset is unexperienced by the expected DeepNet model to appraise how well it achieves underneath different preliminaries.

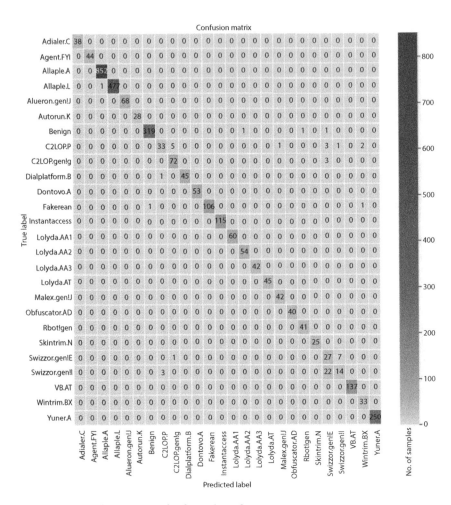

Figure 2.11 Confusion matrix for the Malimg dataset.

The three achieved malware datasets include totally assorted classes from the Malicia dataset gatherings. The assessment of the expected methodologies with the ML and DL approaches across the unnoticed Malicia dataset is coordinated in Table 2.4 and Figure 2.10, respectively. The results on the unnoticed Malicia dataset represent an exactness of 90.2%, which is more prominent than the introductions of the ML and DL approaches across the refined datasets.

The confusion matrices for the models prepared on three malware datasets alongside the harmless class are given in Figures 2.11-2.13. For the

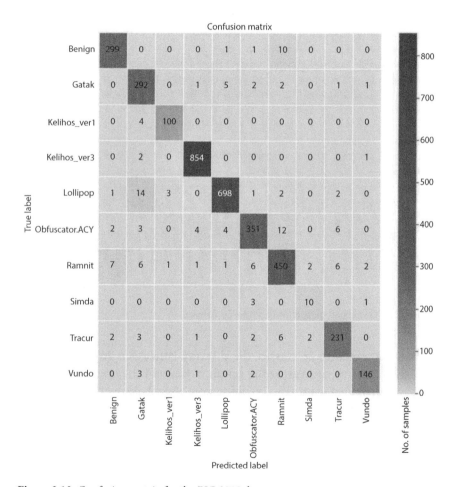

Figure 2.12 Confusion matrix for the BIG 2015 dataset.

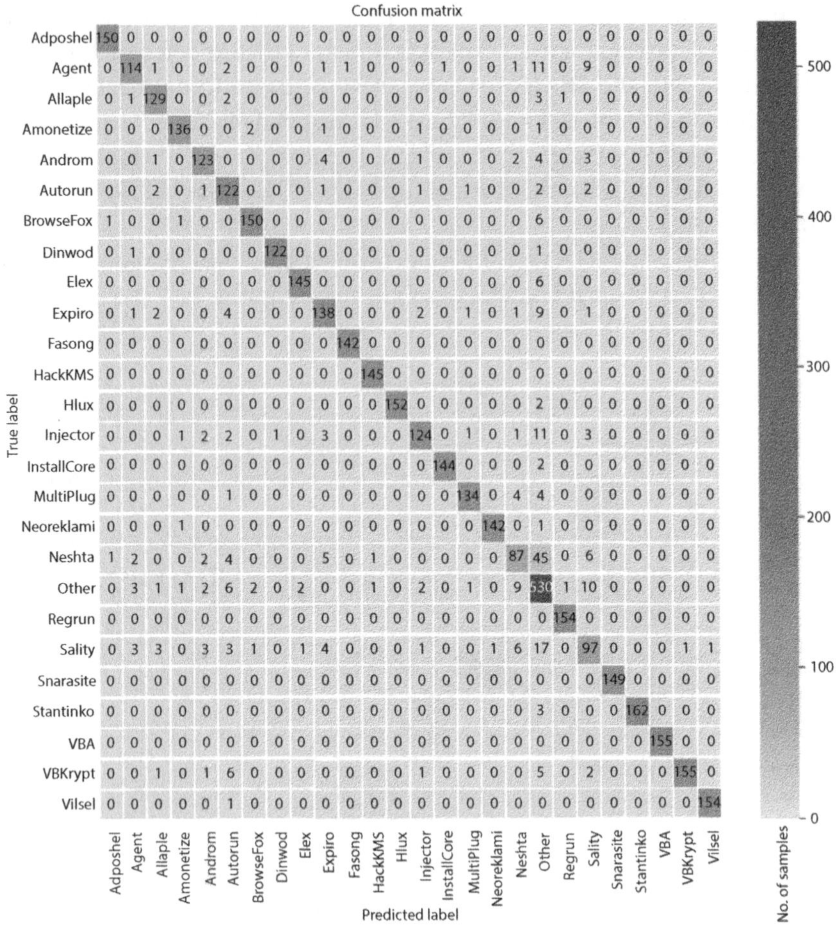

Figure 2.13 Confusion matrix for the MaleVis dataset.

Malimg dataset with 26 classes, the disarray lattice is a 26 x 26 network with the segments addressing the genuine class and the lines demonstrating the anticipated class. The inclining components show the quantity of accurately ordered examples, where the anticipated class matches the real class. The off-inclining components address misclassified tests. The inclining components for each of the three datasets show higher qualities contrasted with the off-corner to corner components. Albeit the examples in the Simda class are less, the vast majority of the examples in that class were accurately ordered by the proposed model.

Table 2.5 gives specifics practically the case held for the expected model to prepare and assess the preliminaries. The assessment of the expected model and the malware identifiers in light of various DL approaches are analyzed in marks of computational viability. The results indicate that the expected DeepNet-based malware acknowledgment model takes less time to prepare and test the assess the malware preliminaries when surveyed to other DL-based malware acknowledgment plans. The time and space complexity for the anticipated DeepNet model is lesser than other existing ML and DL-based models, because the proposed system makes use of balanced and standardized binary input data instead of imbalanced RGB image data as used in the existing ML and DL-based models.

Table 2.6 surveys the results of the expected malware acknowledgment model with going before produces on the four malware datasets (3 preparation dataset + 1 [unnoticed] test dataset). The expected model surpasses other acknowledgment strategies in the connected works. The accuracy of the expected model (98.7%) is possibly more noteworthy than the accuracy of the strategy by Roseline *et al.* (98.6%) on the Malimg dataset. The results of the expected model surpass the overall methodologies on the Enormous 2015, MaleVis, and Malicia datasets.

Table 2.5 Examination of DL-based techniques with the proposed DeepNet model based on computational time.

Models	Training time (seconds)			Testing time (seconds)		
	Malimg	BIG 2015	MaleVis	Malimg	BIG 2015	MaleVis
CNN	6140	4406	10946	7.82	8.14	8.58
VGG16	5174	3652	12721	6.67	6.84	7.04
VGG19	5363	3870	15144	6.35	6.61	6.74
Inception-v3	5604	4146	11379	5.89	6.08	6.36
Resnet-50	6097	4712	8861	7.36	8.12	8.58
Xception	6574	5259	8328	8.48	8.70	8.96
DenseNet-121	5674	4226	10448	5.08	5.53	6.36
DeepNet	**1919**	**2212**	**2391**	**4.40**	**4.50**	**5.10**

Table 2.6 Examination of existing works with the proposed DeepNet model for the four datasets.

Methods	Maling dataset				Big 2015 dataset				MaleVis dataset				Malicia dataset			
	Acc (%)	Pr	Re	F-Score	Acc (%)	Pr	Re	F-score	Acc (%)	Pr	Re	F-score	Acc (%)	Pr	Re	F-score
Nataraj et al. [20]	97.2	0.97	0.97	0.97	96.5	0.96	0.95	0.96	91.7	0.92	0.90	0.91	85.2	0.85	0.85	0.85
Roseline et al [24]	98.6	0.99	0.99	0.99	97.2	0.98	0.97	0.97	97.4	0.98	0.97	0.97	86.5	0.86	0.86	0.86
Cui et al. [28]	94.5	0.95	0.94	0.94	93.4	0.93	0.94	0.93	92.1	0.92	0.92	0.92	80.2	0.79	0.80	0.80
Agarap et al. [29]	84.9	0.85	.85	0.85	80.5	0.81	0.80	0.81	79.4	0.80	0.78	0.80	72.1	0.72	0.72	0.72
Vinayakumar et al. [27]	96.3	0.96	0.96	0.96	91.3	0.92	0.91	0.92	86.3	0.87	0.86	0.87	84.6	0.84	0.84	0.84
Luo et al. [30]	93.7	0.94	0.93	0.93	93.6	0.94	0.93	0.94	92.2	0.92	0.91	0.91	82.5	0.82	0.82	0.82
Singh et al. [31]	96.0	0.96	0.96	0.96	94.2	0.94	0.93	0.94	93	0.93	0.92	0.92	84.3	0.84	0.85	0.84
DeepNet	98.7	0.98	0.98	0.98	98.5	0.99	0.98	0.98	98.2	0.99	0.98	0.98	90.2	0.90	0.90	0.90

2.7 Conclusion & Future Work

There has been broad concentration on malware discovery and grouping yet the capacity to precisely recognize malware variations represents a serious risk to network safety. Malware distinguishing proof is a very troublesome activity because of code disguise and bundling strategies. To precisely recognize malware variations, this work introduced an exceptional profound learning design. The recommended engineering utilizes a half and half methodology. At first, datasets, for example, Malimg, MaleVis and Enormous 2015 were utilized to secure the malware information. The highlights are then recovered utilizing Stacked Auto encoders and Profound Conviction Organizations with Confined Boltzmann Machine. The malware order in the proposed profound brain network design's preparation stage is then done utilizing a softmax classifier.

The essential commitment of the suggested strategy is the introduction of a crossover model made by ideally combining two profound learning designs. The exhibition of the proposed approach is surveyed on the Malicia dataset. The ordinary AI and profound learning models appeared differently in relation to the proposed half breed model. As per the test discoveries, the recommended technique effectively classifies malware with high accuracy, review, precision, and f-score. What is more, it is noticed that the recommended approach is powerful and limits highlight space on a wide space. Second, state-of-the-art strategies were utilized to assess the proposed model. The results gained here likewise uncover and approve the prevalence and benefit of the suggested methodology over other famous courses in the conventional strategies. Then again, a little level of malware preliminaries probably won't be properly distinguished. This is so because such malware renditions portion highlights with other malware classifications and utilize complex code camouflage procedures. Later on, work, a discovery strategy that especially perceives and orders malware that utilization disguise methods might be executed.

References

1. Ö. Aslan and R. Samet, A comprehensive review on malware detection approaches, *IEEE Access*, vol. 8, pp. 6249–6271, Jan. 2020.
2. R. Komatwar and M. Kokare, A survey on malware detection and classification,' *J. Appl. Secur. Res.*, pp. 1–31, Aug. 2020.
3. M. Nisa, J. H. Shah, S. Kanwal, M. Raza, M. A. Khan, R. Damaševičius, and T. Blažauskas, Hybrid malware classification method using segmentation-based

fractal texture analysis and deep convolution neural network features, *Appl. Sci.*, vol. 10, no. 14, p. 4966, 2020.
4. R. Vinayakumar, M. Alazab, K. P. Soman, P. Poornachandran, and S. Venkatraman, Robust intelligent malware detection using deep learning, *IEEE Access*, vol. 7, pp. 46717–46738, 2019.
5. Jeon, J. H. Park, and Y.-S. Jeong, Dynamic analysis for IoT malware detection with convolution neural network model, *IEEE Access*, vol. 8, pp. 96899–96911, 2020.
6. N. Usman, S. Usman, F. Khan, M. A. Jan, A. Sajid, M. Alazab, and P. Watters, Intelligent dynamic malware detection using machine learning in IP reputation for forensics data analytics, *Future Gener. Comput. Syst.*, vol. 118, pp. 124–141, May 2021.
7. M. Alkhateeb and M. Stamp, A dynamic heuristic method for detecting packed malware using naive bayes, in *Proc. Int. Conf. Electr. Comput. Technol. Appl. (ICECTA)*, Nov. 2019, pp. 1–6.
8. Ö. Aslan and A. A. Yilmaz, A New Malware Classification Framework Based on Deep Learning Algorithms, in *IEEE Access*, vol. 9, pp. 87936-87951, 2021, doi: 10.1109/ACCESS.2021.3089586.
9. Mohamed Yunus, Yus Kamalrul & Ngah, Syahrulanuar. (2020). Review of Hybrid Analysis Technique for Malware Detection. *IOP Conference Series: Materials Science and Engineering*. 769. 012075. 10.1088/1757-899X/769/1/012075.
10. Cannarile, Angelo; Dentamaro, Vincenzo; Galantucci, Stefano; Iannacone, Andrea; Impedovo, Donato; and Pirlo, Giuseppe (2022). Comparing Deep Learning and Shallow Learning Techniques for API Calls Malware Prediction: A Study. *Applied Sciences*. 12. 1645. 10.3390/app12031645.
11. S. Panman de Wit, D. Bucur, and J. van der Ham. 2022. Dynamic Detection of Mobile Malware Using Smartphone Data and Machine Learning. *Digital Threats* 3, 2, Article 9 (June 2022), 24 pages. https://doi.org/10.1145/3484246
12. Eduardo de O. Andrade, José Viterbo, Cristina N. Vasconcelos, Joris Guérin, Flavia Cristina Bernardini, A Model Based on LSTM Neural Networks to Identify Five Different Types of Malware, *Procedia Computer Science*, Volume 159, 2019, pp. 182-191, ISSN 1877-0509, https://doi.org/10.1016/j.procs.2019.09.173.
13. Bibi, I., Akhunzada, A., Malik, J., Iqbal, J., Musaddiq, A., & Kim, S. W. (2020). A Dynamic DL-driven architecture to Combat Sophisticated Android Malware. *IEEE Access*, 8, 129600 - 129612. https://doi.org/10.1109/ACCESS.2020.3009819.
14. Martin Kinkead, Stuart Millar, Niall McLaughlin, Philip O'Kane, Towards Explainable CNNs for Android Malware Detection, *Procedia Computer Science*, Volume 184, 2021, pp. 959-965,ISSN 1877-0509, https://doi.org/10.1016/j.procs.2021.03.118.

15. Fei Xiao, Zhaowen Lin, Yi Sun, Yan Ma, Malware Detection Based on Deep Learning of Behavior Graphs, *Mathematical Problems in Engineering*, vol. 2019, Article ID 8195395, 10 pages, 2019. https://doi.org/10.1155/2019/8195395.
16. Hemalatha J, Roseline SA, Geetha S, Kadry S, Damaševičius R. An Efficient DenseNet-Based Deep Learning Model for Malware Detection. *Entropy*. 2021; 23(3):344. https://doi.org/10.3390/e23030344
17. Mahdavifar, Samaneh; Kadir, Andi Fitriah Abdul; Fatemi, Rasool; Alhadidi, Dima; and Ghorbani, Ali. (2020). Dynamic Android Malware Category Classification using Semi-Supervised *Deep Learning*. 515-522. 10.1109/DASC-PICom-CBDCom-CyberSciTech49142.2020.00094.
18. Richard Harang, Ethan M. Rudd, SOREL-20M: A Large Scale Benchmark Dataset for Malicious PE Detection, https://doi.org/10.48550/arXiv.2012.07634.
19. S. Waczak, Artificial neural networks, in *Advanced Methodologies and Technologies in Artificial Intelligence, Computer Simulation and Human-Computer Interaction*. IGI Global, 2019, pp. 40–53.
20. M. Sabokrou, M. Fayyaz, M. Fathy, Z. Moayed, and R. Klette, Deep-anomaly: Fully convolutional neural network for fast anomaly detection in crowded scenes, *Computer Vision and Image Understanding*, vol. 172, pp. 88–97, 2018.
21. Matilda Rhode, Pete Burnap, Kevin Jones, Early-stage malware prediction using recurrent neural networks, *Computers & Security*, Volume 77, 2018, pp. 578-594, ISSN 0167-4048, https://doi.org/10.1016/j.cose.2018.05.010.
22. Nappa, A., Rafique, M. Z., and Caballero, J. (2015). The Malicia dataset: Identification and analysis of drive by download operations. *International Journal of Information Security*, 14(1):15–33.
23. Nataraj, L., Karthikeyan, S., Jacob, G., and Manjunath, B. (2011). Malware images: Visualization and automatic classification. In *Proceedings of the 8th International Symposium on Visualization for Cyber Security*, VizSec '11.
24. Ronen, R.; Radu, M.; Feuerstein, C.; Yom-Tov, E.; Ahmadi, M. Microsoft Malware Classification Challenge. arXiv 2018, arXiv:1802.10135.
25. Bozkir, A.S.; Cankaya, A.O.; Aydos, M. Utilization and Comparison of Convolutional Neural Networks in Malware Recognition. In *Proceedings of the 27th Signal Processing and Communications Applications Conference (SIU), Sivas, Turkey, 24–26 April 2019*; pp. 1–4.
26. Tensorflow. Available online: www.tensorflow.org (accessed on 10 February 2020).
27. Roseline, S.A.; Geetha, S.; Kadry, S.; Nam, Y. Intelligent Vision-based Malware Detection and Classification Using Deep Random Forest Paradigm. *IEEE Access* 2020, 8, 206303–206324.
28. Cui, Z.; Xue, F.; Cai, X.; Cao, Y.; Wang, G.G.; Chen, J. Detection of malicious code variants based on deep learning. *IEEE Trans. Ind. Inform.* 2018, 14, 3187–3196.

29. Agarap, A.F.; Pepito, F.J.H. Towards building an intelligent anti-malware system a deep learning approach using support vector machine (SVM) for malware classification. arXiv 2017, arXiv:1801.00318.
30. Luo, J.S.; Lo, D.C.T. Binary malware image classification using machine learning with local binary pattern. In *Proceedings of the IEEE International Conference on Big Data (Big Data)*, Boston, MA, USA, 11–14 December 2017; pp. 4664–4667.
31. Singh, A.; Handa, A.; Kumar, N.; Shukla, S.K. Malware classification using image representation. In *Proceedings of the International Symposium on Cyber Security Cryptography and Machine Learning, Beer Sheva, Israel, 27–28 June 2019*; pp. 75–92.
32. Cui, Y.; Jia, M.; Lin, T.Y.; Song, Y.; Belongie, S. Class-balanced loss based on effective number of samples. In *Proceedings of the IEEE Conference on Computer Vision and Pattern Recognition, Long Beach, CA, USA, 16–20 June 2019*; pp. 9268–9277.
33. Kingma D, Ba J (2015) Adam: A Method for Stochastic Optimization. *Proceedings of the 3rd International Conference on Learning Representations (ICLR 2015)*. arXiv:1412.6980
34. BengioYoshua and Lamblin Pascal, Greedy layer-wise training of deep networks, in *Advances in Neural Networks*, 2007.
35. Dina Saif, S.M. El-Gokhy, E. Sallam, Deep Belief Networks-based framework for malware detection in Android systems, *Alexandria Engineering Journal*, Volume 57, Issue 4, 2018, Pages 4049-4057, ISSN 1110-0168, https://doi.org/10.1016/j.aej.2018.10.008.
36. Fathima, Nasreen; Pramod, Akshara; Srivastava, Yash; Thomas, Anusha; Syed, Ibrahim S. P.; and Chandran, K. R. Two-stage Deep Stacked Autoencoder with Shallow Learning for Network Intrusion Detection System, 2021.

3
State of Art of Security and Risk in Wireless Environment Along with Healthcare Case Study

Deepa Arora and Oshin Sharma*

Computer Science and Engineering, SRMIST, Delhi-NCR Campus Modinagar, Ghaziabad, UP, India

Abstract

Sending data through a wireless medium is very common these days. The wireless interface is open and available to both authorized and unauthorized users because of the broadcast nature of radio propagation. It is important to ensure that confidential data is only accessed by authorized individuals and not intruders. Security should be the main consideration when deploying wireless networks to monitor applications in hostile unsupervised situations. Wireless security protects a Wi-Fi network from illegal access. It remains difficult to ensure the security of data transfer over wireless networks. At the physical layer of a wireless network, jamming and eavesdropping are two common assaults. Military, sensor node tracking, industry, health applications, home applications, and hybrid networks are the main application areas for wireless networks. All these applications require the security and privacy of data because it contains sensitive data. The most important requirement for healthcare applications is security when it comes to protecting patient privacy. This chapter reviews security risks in wireless networks, including needs, and categories, such as attacks, vulnerabilities, and threats towards healthcare applications.

Keywords: Wireless communication, cyber security, security attacks, security models, active attacks, passive attacks, eavesdropping, security & privacy

---

*Corresponding author: oshins1@srmist.edu.in

3.1 Introduction

A network is the collection of connected computers that allows one node to share resources, information, and programs with another node. A node can be a personal computer, server, hardware, or host. A network may be modest, consisting of only one system, or it may be as vast as desired. The use of a network is for sharing files, maintaining information, accessing the software, using the operating system remotely and resource sharing such as printer, scanner, etc. The medium used for communication can be guided or unguided, and the type of communication system can be wired or wireless. A physical channel such as coaxial cables, twisted wire cables, etc., serves as a medium in wired communication and directs the signals from one point to another. A guided medium is the one that operates in this manner. Examples are wired LAN and Ethernet, etc. Installing a wired network is quite expansive since coaxial cable installation takes a lot of time and money. So, peer to peer technology is currently employed as an option to decrease the costs while simultaneously enhancing networking and reliability. Therefore, wireless networks are installed everywhere instead of those expansive wires.

Wireless communication, on the other hand, does not require a physical channel and instead sends the signals via space. The medium utilized in wireless communication is known as an unguided medium since space only permits unguided signal transmission. A wireless network is an interconnection of computing devices that are not connected by cables. The wireless network is the collection of several networks that enables physical connectivity between computers without the use of wired connections. For connection, generally devices used radio waves. The communication range can vary from a few meters to thousands of kilometres when using radio waves. Being removed from the barrier of a physical network, a wireless network has been used to connect multiple wired organizational structures and to give connectivity within the organization, enabling employees to move around freely. These devices allow roaming within the network coverage and sharing of the information and resources. To provide the mobility feature, the network's topology is constantly changing. Therefore, wireless networks are self-organizing and self-configuring. In addition to transferring data like files and emails, wireless networks are mostly used for audio and video conversions. Wireless environment is very beneficial for real-world applications such as healthcare, industrial, environment monitoring, smart cities, etc. [1].

Figure 3.1 Wireless environment.

Figure 3.1 shows an example of a wireless environment. There are different types of network technologies used according to their network range and performance. Users are allowed to use different network technologies and can switch between them according to their needs. Wireless networks allow users to share information from anywhere within the range of the network topology. A device can be located far away from a router and yet be connected to the network since access points boost Wi-Fi signals. Examples of wireless networks are mobile phone networks, satellite communication networks and wireless sensor networks, 5G Cellular, Wi-Fi, Bluetooth, GPS, etc.

In wireless networks, security will be crucial. Security is a main issue, especially when data is being transmitted between devices and needs to be protected and secure. Even though 3G and 4G networks already have independent security layers, certain well-known types of attacks are still a possibility. Computer network security refers to the steps that organizations take to monitor and stop unwanted access from outside intruders. Network security refers to safety across all networks including network of networks. Steps for computer network security depend on the size of the network employed. For example, a school requires basic security features but a military or banking system requires high security features.

Wireless networks are used in many real-time applications in various fields, such as military and health; as a result, a wireless network requires security to control vital information like personal location, etc. Wireless networks use radio waves rather than wires to transfer the data between devices. Because wireless networks use a broadcast transmission medium, they are vulnerable to security assaults. Wireless networks are suspicious of unauthorized interception, hacking and a variety of cyber threads since they lack physical barriers. Computer network requires security from attackers and hackers. The risk of data modification, removal and theft can be decreased by using a strong network security system. There are two fundamental protections in network security. The first is data security against theft and illegal access. Second is computer security that prevents information from hackers.

3.2 Literature Survey

In most relevant surveys, various facts of wireless network security are investigated. In presenting previous studies on the security of wireless network, this section offers information on some of the most common security risks and challenges with their conclusions.

There are four primary categories of wireless networks: PAN, LAN, MAN, WAN. According to Kanika Sharma *et al.* [2], Wireless personal area networks (WPAN) is the wireless network used to connect devices in a very limited area up to 10 meters. Its range is around a single person in a place. Bluetooth, infrared and zigbee is used for connectivity. Mobile phones, desktop, laptop, tablets, play stations, cordless mice, and headphones are personal devices that are used to create the wireless personal area networks. WPAN are safe networks, restricted to limited coverage range [2]. Examples of wireless personal area network is body area network, smart home appliance, etc. Wireless local area networks (WLAN) are a group of computers and related peripheral that are interconnected in a constrained space, such as school, office, etc. WLAN is the wireless network used to connect devices for short-range communication up to 100 meters. WLAN is also known as Wi-Fi. The coverage range of WLAN is in a limited area such as an office building, healthcare provider, school, hospital or university. It connects different devices such as printers, computers, mobile phones, etc. A small number of users can create a temporary network without an access point. WLAN provides high-speed data transfer rates up to 200 Mbps for a short range. Wireless metropolitan area networks (WMAN) consist of several WLAN. While being smaller than WWAN, a WMAN is larger than a WLAN. The coverage range of WMAN is greater than WLAN and extends to an entire city or geographical up to 50 km. IEEE802.16 standard is used to describe the WMAN. Wireless wide area networks (WWAN) consist of several WMAN. It has a large range, covering several neighbouring cities or states or a country. WAN is also referred to as cellular servicers. Through satellite links, a wireless wide area network expands over a huge geographical area and is not restricted to one site. WAN has multiple Personal networks, local area networks and metropolitan area networks to provide large-area coverage. GSM, GPR are the examples for WWAN. Other previous studies with their findings are given in Table 3.1.

Table 3.1 Previous studies of security risks in wireless environment.

Author	Conclusion
Soo-Hwan Choi et al. [3]	The authors used embedded Bluetooth applications for wireless networks that can benefit from the methodology or algorithm disclosed in this paper.
Kalpana Sharma et al. [4]	According to the authors, due to wireless transmission and resource limitation on wireless sensor network, security designs utilized for conventional wireless networks are not a practical solution to the security issues. Therefore, the nodes are frequently positioned in unsafe conditions in which they are not physically shielded, which makes wireless sensor networks much more vulnerable.
Anitha S. Sastry et al. [5]	In this paper, the authors provide an overview of the numerous threats and security issues in each layer of wireless networks.
Yulong Zou et al. [6]	The authors explain the effective protective mechanisms for enhancing the security of wireless networks; the focus is on physical layer security, and security flaws and threats in a wireless environment are examined.
Yang Gao et al. [7]	In this paper, threats are categorized at physical, network and application layer using the architecture of cyber-physical systems. The authors also provide security breaches of cyber-physical system.
Hiren Kumar Deva Sarma et al. [8]	In this paper the authors identify many security risks that could exist in a wireless network. Mathematical models of the threats have been attempted.
Javier Lopez et al. [9]	The authors explain the summary and evaluation of a connection between the security threats, needs, and uses. They also explain the security requirements of current network standards.

(Continued)

Table 3.1 Previous studies of security risks in wireless environment. (*Continued*)

Author	Conclusion
Aditya Patel *et al.* [10]	According to the authors, wired and wireless networks are becoming more and more vulnerable to a new kind of security threats and flaws, rendering them unreliable and unsafe. They also provide a survey of various security threats and security measures used in an educational system.
Rashid Nazir *et al.* [11]	The security-related problems and difficulties are investigated in wireless networks. The authors also list the potential security risk and consider wireless network security measures.
Al-Sakib Khan Pathan *et al.* [12]	Wireless network technology introduces several security risks. In this paper, the authors explain the comprehensive approach to security that A wireless network should take to provide layered and strong protection.

3.3 Applications of Wireless Networks

Wireless Networks and devices have a wide range of potential applications in human activities. Numerous industries use a wireless network, including those which track animals, monitor traffic, operate connected vehicles, and more. A wireless network is used for a variety of real-time applications as shown in Figure 3.2, such as home healthcare, environmental monitoring, military surveillance, etc., because of its mobility, flexibility, efficiency, easy installation and scalable nature [4]. Some application areas of wireless networks are given below:

Internet access: The most important advantage of a wireless network is having the ability to share a single high-speed internet connection. Wi-Fi and Bluetooth all are because of wireless networks.

Environment monitoring: One of the other main applications of wireless network is environment monitoring. By using environment monitoring we can observe and manage temperature, light, weather, etc. Environment monitoring is used in many different applications such as agriculture monitoring, forest monitoring, habitat monitoring, coal mining, earthquakes, rainfall range, water quality, greenhouse monitoring, climate monitoring, traffic, etc. By using the benefits of wireless network, a environment monitoring system is able to monitor real-time applications.

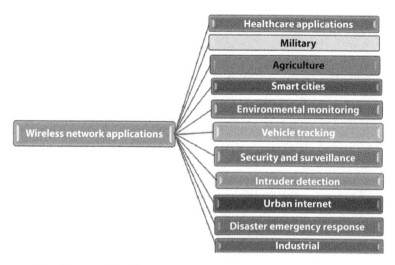

Figure 3.2 Applications of wireless networks.

Healthcare application: Many wireless technologies are applied to health care applications such as sensor networks, RFID, etc. Many healthcare centers implant RFID tags to identify their patients. By using different wearable and implant sensor equipments they can easily monitor and control patient health, such as blood pressure and heart rate. By using mobility a patient can be monitored anytime, anywhere and immediate treatment will be provided in case of emergency.

Education: A wireless network is also useful in the field of education. As we know, during the COVID-19 pandemic all the classes are going on using video communication. We can attend any meeting or online class or seminar from anywhere. Online learning is already widely recognized as the best alternative, allowing for the distribution of knowledge over both time and location.

Industrial applications: Wireless networks are also used in industrial applications for sensing and diagnostics, robotics and machinery health monitoring. It is used in a variety of industrial applications to address a wide range of connected issues. Wireless network application for logistics make use of GPS technology. This system uses an embedded terminal to find the items and a cloud service platform to identify the recipient to monitor the status of the goods in real time. The development of wireless network enables monitoring of electric machine status and energy usage.

Smart homes: Wireless networks are used in indoor environments such as in smart homes, where machine-to-machine connections take place. A smart home can easily operate home appliances such as lights, CCTV cameras, child monitoring system remotely from mobile phones by using wireless networks. This allows us to operate these devices anytime, anywhere by using mobile phones. Motion monitoring indoors and indoor air quality monitoring are two common examples of smart homes.

Connected vehicles: A vehicle that can connect to equipment nearby or far away using wireless network is referred to as a connected vehicle. For location tracking this is mostly used in vehicles. By using GPS, we can easily find the location of the vehicle.

3.4 Types of Attacks

There are two types of attacks in a wireless environment: passive attacks and active attacks. These are explained below.

3.4.1 Passive Attacks

In passive attacks the attacker tries to learn something or obtain information. These attacks do not modify or remove the data. The hacker only captures the data during transition [13]. Figure 3.3 shows types of passive attacks, which are explained below.

3.4.2 Release of Message Contents

In this type of attack, the hacker obtains the data without the permission of sender and receiver of the communication system. For example, if the sender sends an email to the receiver and a hacker obtains the information from that email and sends it to someone else without permission of sender and receiver.

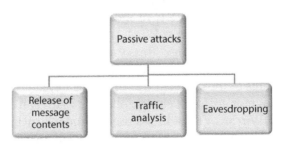

Figure 3.3 Types of passive attacks.

Prevention: This type of attack can be handled by encryption so that the hacker cannot easily learn from the data.

3.4.3 Traffic Analysis

Suppose we decode the data by using encryption, the hacker may capture the information but cannot obtain the data from the message. In this type of passive attack, the hacker discovers the pattern of data flow during communication.
Prevention: use of strong encryption algorithms and masking.

3.4.4 Eavesdropping

Eavesdropping is also known as snooping. In this, someone listens to a secret conversation between sender and receiver. The result of eavesdropping is that a hacker can intercept, remove, or modify the data between devices. An example of eavesdropping is to listen to a quarrel between your neighbours through a vent in your apartment.
Prevention: VPN is used to prevent eavesdropping.

3.5 Active Attacks

In active attacks, the attacker not only steals the data but also modifies or deletes the message. Systems can be harmed by these attacks [11]. There are various types of active attacks, which are shown in Figure 3.4.

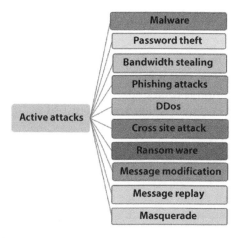

Figure 3.4 Types of active attacks.

3.5.1 Malware

Malware occurs when an unauthorized programme or piece of software gains access to a target system and exhibits strange behaviour. The effects of malware are to remove important files, capture the information and not provide access to programs.
Prevention: for the prevention of malware a proactive approach is used. Anti-malware programs and antivirus are used to detect the malware.

3.5.2 Password Theft

This is another security risk in a wireless environment. In password theft someone guesses or steals the password and the result will be the loss of information.
Prevention: there are two ways to prevent password theft. One method is robust protection which requires an additional device to login, such as login is possible only with confirmation done by mobile. Another method is to use complicated logins to avoid brute force.

3.5.3 Bandwidth Stealing

A wireless environment is available for outer intruders also. They can lower the speed of the network by downloading games, music, etc., over the internet connections.
Prevention: Limit the bandwidth according to number of employees to whom you want to provide access.

3.5.4 Phishing Attacks

Phishing attacks are very common nowadays. The hacker sends a link or attachment which requires sensitive data such as a password and compels the end user to click on that link.
Prevention: Precaution is used to prevent phishing attacks. Official emails from organizations do not require your password so try to avoid filling in a password on the unknown links.

3.5.5 DDoS

DDoS is distributed denial of service attack in which a hacker sends many requests to target servers so that the server cannot handle these requests and slows down or crypts the server. In this type of attack, the server is overloaded or slowed down by sending a number of false data requests to the server. This will cause problems for authorized users of the server to use the services of the server.

Prevention: prevention of DDoS is by detecting suspicious traffic and placing the server offline for maintenance.

3.5.6 Cross-Site Attack

In a cross-site attack the attacker loads dangerous codes into a website. The goal of this attack is to steal information or disturb standard services.

Prevention: Stopping a cross-site attack is still a challenge. It depends on the website owner's ability to find it and fix it.

3.5.7 Ransomware

As we know, data today is very important and all the data has been stored on a system so ransomware is the malware that installs itself on the system and after that it will not give access to either the whole system or a particular part of it. To provide access to that data the hacker asks for some ransom amount. The hacker encrypts the computer system and demands an amount to decrypt them. The target of these attacks is the systems or organization for which paying ransom is easy in order to regain the data. An example is a banking system. The data in a banking system is very precious so if the workers in a bank do not have access to the data of users it creates a serious problem.

Prevention: To stop this malware after installation is very difficult; there is only precaution to avoid this. Best prevention of this attack is to have updated antivirus and avoid any suspicious link. Backups made from time to time and replications of data are used to prevent this.

3.5.8 Message Modification

In this type of active attack, the attacker modifies the message during communication. The attacker captures the message sent by the sender and then modifies it and sends the modified message to the receiver.

Prevention: To stop this type of attack, use strong encryption algorithms.

3.5.9 Message Replay

The attacker captures the message during transmission and then replays or retransmits this message to the receiver so that the receiver receives the same message multiple times.

Prevention: Encryption is used to prevent this attack by encrypting the transmission between sender and receiver. By using scrambling, the server makes your data unreadable to an outsider.

3.5.10 Masquerade

In this type of active attack, the attacker pretends to be another authorized person. Using masquerade, the attacker wants to access the data of that authorized person.

Prevention: By using authentication, we can stop this type of attack.

3.6 Layered Attacks in WSN

The open systems interconnection (OSI) model was developed by the International Organization for Standardization (ISO) in 1984. This model has seven layers and each layer has a specific task to perform. The computer systems employ these seven layers to interact over a network [5]. This section explains the attacks in different layers of OSI model which are shown in Table 3.2.

Table 3.2 Security attacks in OSI protocol layers.

Layers	Attacks
Application layer	Malware attack, SQL injection, cross-site scripting, FTP bounce, SMTP attack, attacks on reliability.
Transport layer	TCP flooding, UDP flooding, desynchronization, TCP sequence andprediction attack, data integrity, energy drain attack.
Network layer	Neglect and greed, homing, misdirection, hello flood attack, black holes, spoofing, sink holes, IP hijacking, Sybil attacks, node replications, worm holes, flooding, attack against privacy, internet smurf attack [12].
Data link layer	Collision, jamming, exhaustion, interrogation attack, Sybil attack, data aggregation, voting , MAC spoofing, identity theft, MITMattack, MAC flooding, unfairness [12].
Physical layer	Eavesdropping, jamming, tampering, side channel attack, Sybil attack, random interference and timing attack [8].

3.6.1 Attacks in Physical Layer

Eavesdropping: An attack in which a hacker listens to personal data without knowledge or permission of sender and receiver of the system.

Jamming: Jamming radio signals transmission causes a problem with WSN's radio frequencies. Therefore, the transmission of data is stopped.

Tampering: The attacker intentionally modifies the data in a way that would be harmful for the users.

Side channel attack: Depending on the physical properties of a cryptosystem, the attacker finds the secret information.

Sybil attack: The attacker creates multiple identities to slow down the network speed.

Random inference: The hacker randomly interrupts the user of the communication system.

Timing attack: This attack involves a calculation of time to perform encryption or decryption to obtain a key.

3.6.2 Attacks in Data Link Layer

Collision: This occurs if a channel is occupied by another sensor node, and therefore a lot of data is lost due to collision.

Jamming: This happens when a radio frequency from the other broadcasts interferes with the data transmission.

MAC spoofing: It is also referred to as counterfeiting of MAC address. In order to access wireless networks, MAC spoofing is frequently utilized. The attacker attacks a network to obtain valid MAC address and modify the media access control address.

Identity theft: When someone steals your personal information, identity theft occurs. It can be done in multiple ways. In data link layer, the attacker steals the MAC address of user.

Man in the middle (MITM) attack: This occurs when an attacker interferes with the user's interaction with an application.

MAC flooding: It is a technique to determine the security of network switches.

3.6.3 Attacks in Network Layer

Hello flood attack: The attacker sends hello packets to sensor node to pretend that this is neighbour node and try to get the data packets.
Black holes: The attacker pretends to have the shortest path from node to base station and if the sensor node chooses that path, then malicious node hack the data.
IP Spoofing: It is falsification of IP address.
Sinkholes: The attacker attempts to create a lot of traffic in base station to interrupt the sensing data coming from the nodes.
IP hijacking: In this type of attack, the attacker hijacks or steals the IP addresses.
Wormholes: In a wormholes attack, a malicious party replays messages that have been received in one area of the network through a low latency channel.
Flooding: In flooding, an attacker sends multiple data packets to slow down the network.
Attack against privacy: The attacker steals the personal information of the users through the network.
Internet smurf attack: The attacker sends multiple ICMP requests to halt the network.

3.6.4 Attacks in Transport Layer

TCP flooding: This is referred to as part of DDoS, also known as SYN flood; it takes advantage of a portion of the typical TCP three-way handshake to deplete the resources of the server and make it unavailable.
UDP flooding: This is also referred to as part of DDoS; the attacker floods the targeted host's random ports with IP packets including UDP datagram.
Desynchronization: This is also referred to as TCP hijacking. It's a procedure where the expected sequential number and the sequential number in a received packet are different.
TCP sequence and prediction attack: An attacker predicts TCP sequence number for the creation of a legal user data package.
Data integrity: Data integrity attacks alter or introduce fake data into packets, which determine the data being transmitted between WSN nodes.
Energy drain attack: As we know, in a wireless network, sensor nodes have limited battery power. So the attacker sends a false alarm which drains the battery power of sensor nodes.

3.6.5 Attacks in Application Layer

Malware attack: Malicious software is produced by an attacker in the form of programming, scripting and data.
SQL injection: In this attack, an attacker uses a rogue SQL statement to get unwanted access to trustworthy websites.
Cross-site scripting: The attacker tries to get a number of access control measures by adding client side scripts to websites.
FTP bounce: An attacker sends unbound traffic to another server of the network to get unauthorized access.
SMTP attack: Threats to the transmission of SMTP server and client emails.
Attacks on reliability: The attacker attempts to obtain the communication path by sending a false query. A node will experience energy drain when it responds to this false query.

3.7 Security Models

There are many methods to enhance the security in a wireless environment such as trust and reputation security models, secure routing protocols and intrusion detection systems. Trust and reputation security models improve the security in wireless environments as explained below [14].

3.7.1 Bio-Inspired Trust and Reputation Model

The most reliable node along the most reliable path providing a specific service is chosen by BTRM-WSN [14]. It is based on the Ant Colony Systems (ACS), a bio-inspired algorithm based on ants' construct routes to graphically satisfy certain constraints. The ants leave behind some pheromone traces that aid other ants in locating and travelling along the same paths. Ants will use these pheromone values to determine the best routes because the best path will have the highest concentration of pheromone value. We utilize "pheromone value" to represent the credibility of sensors when we apply our ACS algorithm to a trust and reputation system. Each sensor carries pheromone traces for its neighbours, determining the likelihood that an ant would choose a path. Artificial ants are constructed, and they eventually depart from the client sensor. When an ant moves from one sensor to sensor, it sends an instruction to these sensors via equation 3.1 and equation 3.2 to change the pheromone value of the route between them.

$$\tau ij = (1- \varphi). \tau ij + \varphi.\Omega \quad (3.1)$$

$$\Omega = (1+ (1- \varphi). (1- \tau ij).\eta ij \quad (3.2)$$

There are various scenarios that could happen when an ant k gets to sensor s. The first scenario is when sensor s provides the service then the average phenomenon value of the route taken by ant k from the client until the sensor s is calculated $\tau[0,1]$. If the sensor s has more neighbours that have not been visited by ant k, ant k pauses and returns if exceeds the predetermined transition threshold (TraTh) and vice versa. Ant k stops and returns the solution if sensor s has no more neighbours or if every one of them has already been visited. Another scenario is when a sensor s does not offer any services. If sensor s still has neighbours that ant k has not visited yet, k chooses which node to travel next. Ant k runs into trouble if sensor s has no more neighbours or if each one has already been explored. It must retrace its steps until it reaches either:

a) A sensor that provides the required service.
b) A sensor that does not, but has other nearby sensors that have not been visited yet.

The client will look over and evaluate the calibre of each launched ant's response. Equation 3.3 is used to calculate the route quality.

$$Q(Sk) = (\overline{\tau k}/ \text{length } (S_K)^{PLF}).\%A_k \quad (3.3)$$

Where, φ is the parameter controlling how much pheromone the ants leave behind, τij is the pheromone value of the route between sensors i and j, Convergence value of τij is Ω,
S_k Is the solution given back by ant k. $Q(S_k)$ is the quality of the selected route.
$\overline{\tau k}$ is the average path pheromone of route S_k, Percentage of ants used for the solution as ant k is $\%A_k$, And path length factor is PLF.

3.7.2 Peer Trust System

The basic goal of dynamic peer-to-peer trust and reputation model, known as the peer trust model, is to estimate and assess a peer's reliability or quality, in an online commercial context [14]. For calculating the reliability value of a given peer, it identifies five trust and reputation management related factors: 1) the responses a peer receives from others; 2) the response scope or field; 3) the source creditability; 4) the transaction scenario variable designed to address the essentialness of transaction; 5) the community scenario variable trying to interpreting related features. In wireless networks equation 3.4 could be used to determine the trust value of peer u.

$$T(u) = \alpha \cdot \Sigma\, S(u, i) \cdot CR(p(u, i)) \cdot TF(u, i) + \beta \cdot CF(u) \quad (3.4)$$

Where, $T(u)$ is the value of reliability of peer u.
α is the weight factor used for evaluation
β is the weight factor used for community scenario variable
And $S(u,i)$ is the normalized amount of satisfaction which peer u received in i transaction.

Therefore, these are two security models by which we can enhance the security in wireless networks.

3.8 Case Study: Healthcare

Wireless communication is beneficial for real-time applications such as in entertainment, transport, shopping, industry, medical and many other areas. Wireless network topology has the potential to revolutionize the way we live. The most important application of wireless network is healthcare. It is referred to as wireless medical sensor network (WMSN). The main concern of WMSN is patient mobility and reliable communication. The creation of wireless healthcare application presents many issues including timely distribution of data, quick event detection, power management, etc.

Figure 3.5 shows the healthcare system. In this figure, multiple sensors are attached to the human body. These sensors can be a wearable device such as a smart watch or implant in the human body. Sensors are connected to a cell phone or gateway by using a wireless network and these are connected to the internet and send information to different recipients, for example to call an ambulance or inform the hospital and family members or take a prescription to a specialist doctor and immediate treatment will be provided in case of emergency. Later, by using the server this information can be saved to the database.

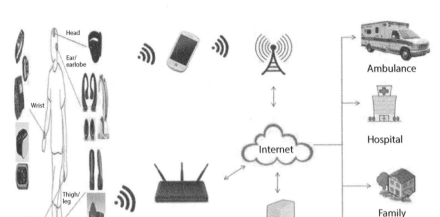

Figure 3.5 Application of wireless network towards healthcare.

3.8.1 Security Risks in Healthcare

Hospitals and other healthcare facilities face many active attack and passive attacks [15]. As we know, in a passive attack the attacker steals the information but does not modify or delete it. Therefore, in healthcare many attackers capture the confidential report but do not modify it. These days, one of the most popular active attacks is ransomware, which affect confidentiality. According to the survey, the rate of healthcare ransomware attacks is rising, making the industry more vulnerable to a wide range of threats [16]. In this attack the hacker locks the whole system or part of the system and asks for a ransom amount to reopen the patient-related sensitive and confidential data. According to a report [17] "The number of ransomware attacks on US healthcare organization increased 94% from 2021-2022, according to one report". Healthcare centres are a regular target for ransomware because they rely so much on access to data, such as patient information, to keep their operations running smoothly.

Figure 3.6 shows the ransomware attack cycle. As the figure explains, the first attacker or hacker send a malicious code or link through a phishing e-mail to the target system. After clicking on that link or malicious code the execution of malicious code and the searching of important file

Security & Risks in Wireless Environment: Healthcare Case Study

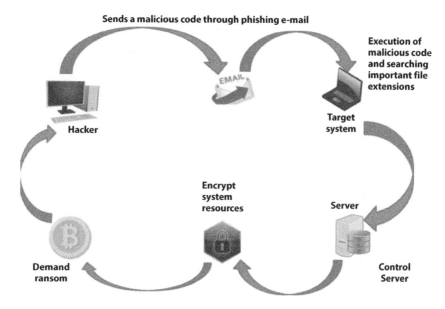

Figure 3.6 Ransomware attack in healthcare.

extensions are started. After this process, the attacker can command and control the server and encrypt the computer system resources. Encryption can be applied to the whole system or in part of the system. Therefore, an attacker demands a ransom amount; most commonly it is bitcoin in the form of ransom. When the amount is sent to the attacker, the attacker decrypt or unlock the system.

3.8.2 Prevention from Security Attacks in Healthcare

The prevention of passive attacks is one use of a firewall. Try to have updated antivirus. For ransomware, there is only one precaution to take. Avoid clicking on a suspicious link because the hacker will send a malicious link via email or through advertisement to hack the system. Create timely backups and make redundancies of data to prevent from this attack. The majority of healthcare centres opt to purchase cyber insurance to minimize the financial risk involved with such an assault. Some steps can be taken if you are facing a ransomware attack, as shown in Figure 3.7.

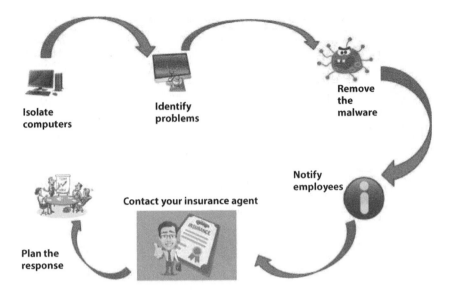

Figure 3.7 Steps to do after ransomware attack in wireless networks.

Step 1: Isolate the computers so that this malware does not infect the other systems.
Step 2: Try to identify the malware or malicious link.
Step 3: Once you identify that malware, remove it.
Step 4: Notify the employees.
Step 5: Contact your insurance agent regarding this.

3.9 Minimize the Risks in a Wireless Environment

Figure 3.8 shows steps to avoid risks in a wireless environment. You can follow these easy procedures to secure your wireless network [18]. These steps are given below:

3.9.1 Generate Strong Passwords

Generate strong passwords with the combination of minimum eight character and having one large alphabet, numbers, and special characters.

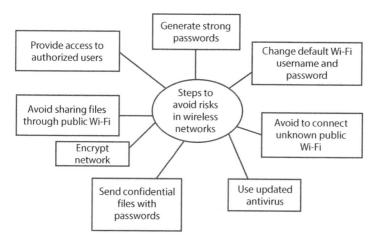

Figure 3.8 Steps to avoid risks in wireless environment.

3.9.2 Change Default Wi-Fi Username and Password

Change default Wi-Fi username and password by giving your SSID a new name, altering its default settings, and turning off its broadcast to users. Avoid connecting unknown open public Wi-Fi.

3.9.3 Use Updated Antivirus

Try to use updated antivirus software. Place a host-based firewall, make sure your security policy is robust and apply the policy of the rule-based firewall configuration.

3.9.4 Send Confidential Files with Passwords

While transferring the data or files use a password to open them. For example, for banking KYC e-mail they provide a pdf having a password that is the DOB of that person so that only authorized person can see the data.

3.9.5 Detect the Intruders

Install an intrusion detection system based on the network, and analyze the log weekly and use updated antivirus software. Updates for anti-virus software are distributed from servers to clients. Make frequent data backups and restore data as needed.

3.9.6 Encrypt Network

Encryption is used to protect the data in the wireless network. Encrypt all WLAN traffic at all times. Application encryption programmers like pretty good privacy; secure socket layer should be used so that the attacker cannot easily hack the data. Current standards for encryption WAP3 to encrypt your data are advisable to use. Encryption should be applied to protect wireless network traffic to avoid traffic analysis attack. If the wireless network is not being used for a long time, turn it off so it will prevent hackers. If the network is not secure do not open any important data or enter credit card information.

3.9.7 Avoid Sharing Files Through Public Wi-Fi

Avoid sharing important files by using public Wi-Fi. Mostly public Wi-Fi are not safe to share private information. Restrict the access of wireless network to only authorized persons.

3.9.8 Provide Access to Authorized Users

Only authorized users should be allowed to access. Create a guest account with visitors' permissions on a different wireless channel; for instance, a guest is required to access the network in order to protect the confidentiality of primary credentials. This concept would be beneficial to ensure that employee and visitor traffic is routed through different network channels.

3.9.9 Used a Wireless Controller

A wireless controller is a gadget that coordinates the provisioning, operating and management of access points. As the access points register to this controller, it will be possible to configure and operate the entire wireless network from the controller's interface as a single entity.

3.10 Conclusion

In this chapter we present a study of security risks in wireless networks defensive techniques to defend the availability, confidentiality, and integrity of the network against malicious intrusions. Commercial applications, as well as the public and private sectors, are gradually using wireless networks. Sensor networks are in high demand for real-world settings;

therefore, it is necessary to provide efficient and usable security mechanisms that could protect the network against attacks. Most security issues in a wireless network are due to insertion of fake information by malicious nodes. The network becomes vulnerable because access points or wireless devices are installed in a public environment. It is very easy to interfere with a wireless network and challenging to prevent it. This chapter describes the security attacks in protocol stack from physical layer to application layer and security models to minimize the risks in wireless networks. Moreover, this chapter provides insights about a healthcare case study along with its risks and preventions.

References

[1] S. K. Mazumder, *Wireless networking based control.* Springer New York, 2011. doi: 10.1007/978-1-4419-7393-1.

[2] K. Sharma and N. Dhir, "A Study of Wireless Networks: WLANs, WPANs, WMANs, and WWANs with Comparison," *Int. J. Comput. Sci. Inf. Technol.*, vol. 5, no. 6, pp. 7810–7813, 2014, [Online]. Available: www.ijcsit.com

[3] S. H. Choi, B. K. Kim, J. Park, C. H. Kang, and D. S. Eom, "An implementation of wireless sensor network for security system using Bluetooth," *IEEE Trans. Consum. Electron.*, vol. 50, no. 1, pp. 236–244, 2004, doi: 10.1109/TCE.2004.1277868.

[4] K. Sharma and M. Ghose, "Wireless Sensor Networks: An Overview on its Security Threats," *IJCA*, vol. Special Is, no. Mobile Ad-hoc Networks" *MANETs*, pp. 42–45, 2010.

[5] A. S. Sastry, S. Sulthana, and S. Vagdevi, "Security Threats in Wireless Sensor Networks in Each Layer," *Int. J. Adv. Netw. Appl.*, vol. 4, no. 4, pp. 1657–1661, 2013.

[6] Y. Zou, J. Zhu, X. Wang, and L. Hanzo, "A Survey on Wireless Security: Technical Challenges, Recent Advances, and Future Trends," *Proc. IEEE*, vol. 104, no. 9, pp. 1727–1765, 2016, doi: 10.1109/JPROC.2016.2558521.

[7] Y. Gao et al., "Analysis of security threats and vulnerability for cyber-physical systems," *Proc. 2013 3rd Int. Conf. Comput. Sci. Netw. Technol. ICCSNT 2013*, pp. 50–55, 2014, doi: 10.1109/ICCSNT.2013.6967062.

[8] H. Sarma and A. Kar, "Security Threats in Wireless Sensor Networks," *IEEE*, pp. 243–251, 2006.

[9] J. Lopez, R. Roman, and C. Alcaraz, "Analysis of security threats, requirements, technologies and standards in ireless sensor networks," *Lect. Notes Comput. Sci. (including Subser. Lect. Notes Artif. Intell. Lect. Notes Bioinformatics)*, vol. 5705 LNCS, pp. 289–338, 2009, doi: 10.1007/978-3-642-03829-7_10.

[10] A. Patel, S. Ghaghda, and P. Nagecha, "Model for security in wired and wireless network for education," *2014 Int. Conf. Comput. Sustain. Glob. Dev. INDIACom 2014*, pp. 699–704, 2014, doi: 10.1109/IndiaCom.2014.6828051.
[11] R. Nazir, A. A. laghari, K. Kumar, S. David, and M. Ali, "Survey on Wireless Network Security," *Arch. Comput. Methods Eng.*, vol. 29, no. 3, pp. 1591–1610, May 2022, doi: 10.1007/s11831-021-09631-5.
[12] A.-S. Pathan, H. Lee, and C. Hong, "Security in Wireless Sensor Networks: Issues and Challenges," *ICACT*, pp. 1043–1048, 2006.
[13] A. Gupta and R. K. Jha, "Security threats of wireless networks: A survey," in *International Conference on Computing, Communication and Automation, ICCCA 2015*, Jul. 2015, pp. 389–395. doi: 10.1109/CCAA.2015.7148407.
[14] H. Marzi and A. Marzi, "A security model for wireless sensor networks," *CIVEMSA 2014 - 2014 IEEE Conf. Comput. Intell. Virtual Environ. Meas. Syst. Appl. Proc.*, pp. 64–69, 2014, doi: 10.1109/CIVEMSA.2014.6841440.
[15] P. Mahendru, "The State of Ransomware in Healthcare 2022," 2022.
[16] M. Berry, "Ransomware attacks against healthcare organizations nearly doubled in 2021, report says," 2022.
[17] K. Paul, "'Lives are at stake': hacking of US hospitals highlights deadly risk of ransomware," 2022.
[18] A. N. Kadhim and S. B. Sadkhan, "Security Threats in Wireless Network Communication-Status, Challenges, and Future Trends," in *2021 International Conference on Advanced Computer Applications, ACA 2021*, 2021, pp. 176–181. doi: 10.1109/ACA52198.2021.9626810.

4
Machine Learning-Based Malicious Threat Detection and Security Analysis on Software-Defined Networking for Industry 4.0

J. Ramprasath[1*], N. Praveen Sundra Kumar[1], N. Krishnaraj[2] and M. Gomathi[3]

[1]Deparment of Information Technology, Dr. Mahalingam College of Engineering and Technology, Pollachi, India
[2]School of Computer Science and Engineering, Vellore Institute of Technology, Vellore, India
[3]Department of Computer Science and Engineering, Dr. Mahalingam College of Engineering and Technology, Pollachi, India

Abstract

Traditional data communication networking is not suitable for industry 4.0. So industry moves to implement software-defined networks for managing the networks. But security is an important thread in a software-defined network. Most attackers are easily getting access to the resource in the industry. The Denial of Service (DoS) and Distributed Denial of Service (DDoS) attacks cause heavy damage in the production area. Malicious attacks will create a services gap; network services throughput enter into a down state, and there is a loss in business continuity. Traditional Intrusion Detection System (IDS) will detect malicious traffic based on a predefined access control list but it cannot detect new malicious traffic ingress into home networks. Machine learning techniques will lead to better identification of threats to synthetic or real-time data. To avoid these situations, we are proposing a model to find the attackers in the network and train the model to find the new type of intruder in the system.

***Keywords*:** Malicious traffic, software defined networking, intrusion detection system, machine learning, traffic analysis

Corresponding author: jrprasath@gmail.com

4.1 Introduction

The fundamentals of networks have not changed much during the past ten years, despite substantial advancements in computer sciences. The ever-evolving needs of the informatics industry and the expanding and emerging data centers have demonstrated that the conventional approach to network administration is no longer adequate. The rapid development of technologies uses the Internet of Things to access everything in the computation field [18]. Most people use internet for developing their business, so network administration needs to be more active and amenable to mixing with novel structures. Applications using distributed architectural structures and the variety of users using new generation devices opened the way for the requirement of readily maintaining networks without the need for people support. A method of network management which doesn't take into account these factors could lead to confusion about the already constantly evolving and constructing services.

4.1.1 Software-Defined Network

SDN has adopted a new architecture and conventional networks' position because of its capacity for responding quickly and easily to new events. In a software-defined network there is no need to monitor each and every device in the network. Since network control is now directly programmable, network infrastructure components like switches and routers are now separated from network services. Network control and forwarding operations are separated in SDN architecture [12].

In the software-defined network the control plane and information plane are separated for the efficient usage of the components. The organizer plays a vital role in managing and guiding each and every switch in the network; in a software-defined network the controller plane will manage and monitor all the nodes in the network. On the basis of the configurations provided by the organizer plane, the information plane is the network architectural layer that physically manages the traffic. The data transmission layer is nothing more than a group of switches that are interconnected. These switches are in charge of acting on the received packet in accordance with the flow rules listed in the flow table. The switch keeps each new packet that enters it in a buffer. The availability of the buffered packet in the flow table is then verified. If there isn't a current rule for that particular packet in the flow table, the packet-IN message is forwarded to the organizer to create one. The modification's flow table is then updated with this new flow rule. Because the switch's buffer and flow table have a limited amount of memory, they are vulnerable to DoS attacks. Given the vulnerability of the shift's limited Drift Table warehouse size, a hacker transport spoof packets with internet protocol bluffing will have to add a novel regulation to be injected in the switch. As a result, a hacker can quickly produce a large number of packets and send them to unnamed network hosts, saturating the buffer and rapidly substantial the Movement Board. Unfortunately, this will result in usual, honest circulation not being routed through the switch [1].

4.1.2 Types of Attacks

Denial of Services (DoS)

- Volume-based assaults
- Protocol assaults
- Application-layer assaults
- UDP Flood
- Internet Control Message Protocol Flood
- Chink of death
- Slowloris
- Enlargement NTP
- HTTP flood

Distributed Denial of Services (DDoS)

- Application layer assaults
- Protocol assaults

4.1.2.1 Denial of Services

The DoS (Denial of Service) attack is the most prevalent type of security risk for networked systems. By monopolizing network resources, such as a server, it seeks to prevent intended users from using them. Sending IP packets to a victim in order to produce saturation or instability is the basic idea behind a DoS attack. Due to their evolution into increasingly sophisticated and varied DoS attacks, it is becoming more difficult to detect them. Furthermore, even without any technical expertise, anyone with attack tools may carry out DoS attacks with ease. Consequently, it still poses a serious hazard. However, because the headers of attack packets are

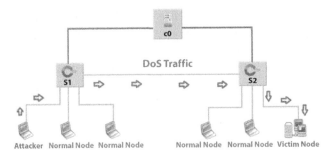

Figure 4.1 Operation of denial of service.

fabricated to seem like normal ones, it is difficult to distinguish between normal and attack traffic [14].

In the denial of service an attacker floods the packets in a network in order to create trouble for the software network as shown in Figure 4.1. Flooded packets create overloaded services to the network and reduce its service. Another important impact of the DoS attack is to deny the needed website to the user at a time of emergency. Packet injection is another type of attack which reduces the performance of the network. The flooded packets will create anonymous traffic delay in the network which creates an artificial delay to the user. The user believes that the delay in the traffic process is because so many communications happen in the network [2].

Volume-based assaults
The key goal of the attacker is to reduce the bandwidth of the site in minutes per second. This form of attacks contains spoof-packet floods, ICMP floods and UDP floods.

Protocol Assaults
The main purpose of this attack is to utilize the honest server resources and its components used for communication, load balancing and firewalls implementation. The broadcast rate is calculated by packers per second. The different category of attacks like Chink of Death, Smurf denial of service, SYN floods, and fragmented packet attack fall in this type.

Application-Layer Assaults
This type of assault, which is measured in requests per secaim, aims to take down the web server. It targets certain platforms like Apache, OpenBSD, and Windows. Two examples of the attacks are

- GROW/STAKE floods
- Low-and-Slow attacks.

UDP Flood
The full attention of the UDP assaults is to overflow accidental ports on the remote host; it is subject to UDP flood attacks. The host keeps looking for the application ports, and if none are found, it responds with an ICMP message indicating that the destination is inaccessible. This has an impact on the host resources and makes services unavailable. User Datagram Protocol (UDP) packets are used to affect and attack the host, as the name suggests.

Internet Control Message Protocol Attacks
In light of the fact that all vulnerable servers try to respond with ICMP repeat response packets constantly, which causes the system to crash or slow down, ICMP assaults use both arriving and departing bandwidth. It is comparable to UDP assaults but instead of waiting for a response, it sends ICMP echo request packets at a high transmission rate to the target. The requestor transmits numerous SYN requests during a SYN flood attack, but never responds to the host's SYN-ACK response, or it transmits the SYN request from a fake or hidden IP address. It is now necessary for the host server to wait for each request to be acknowledged by the receiver and for the permanent required of properties up until the founding of new connections, which finally leads to DoS. This one occurs in order to exploit a known weakness in the TCP joining sequence. It looks like a three-hand salute. Any SYN request that must be sent across a TCP connection to one or more host servers must first be acknowledged with SYN-ACK answers before being verified by the requestor with ACK messages [25, 26]. These assaults therefore affect the service requestor's response.

Chink of Death
This kind of violence entails transfer the server a stable stream of malicious or malfunctioning rings. The IP packet's extreme length, with the header, is 65,535 bytes. The maximum frame size allowed by the data connection layer over Ethernet is 1,500 bytes. A supreme Internet protocol packet is distributed into several Internet protocol rubbles in this circumstance, and the receiving host has the necessary IP packets or fragments to complete the IP [27, 28]. The receiver packets produced when the malware reassembles the fragmented data are larger than 65,535 bytes. Even genuine and authentic packets may experience denial of service if the memory space allotted for the packet is exceeded [15].

Slowloris

This kind of attack has a significant effect, such as permitting one network server while shutting down the other server without affecting other host network ports or facilities [13]. It accomplishes this by beginning connections to the host server but only transmitting incomplete demands, holding numerous networks to the host web server for as extended as promising. It consistently sends more HTTP headers, but it never fulfils the demand. The cloud method keeps the port or facilities open for this fictitious association, which reduces the available capacity for valid requests. As implied by the name, this slows down the entire system by exceeding the allowable number of concurrent connections [19].

Enlargement of NTP

In this kind of assault, the attacker targets users of System Period Rules in order to overwhelm a host server with UDP flood. The term "amplification stabbing" refers to situations where the proportion of a demand to a reaction is substantially higher than 1:100 or 1:220. It means that the hacker has access to a list of available NTP servers and can launch DoS assaults with the highest possible volume and distressing maximum bandwidth. Only NTP protocols are the focus of this kind of assault [17].

HTTP Flood

In this instance, the hacker targets the standard and lawful HTTP GROW or STAKE reaction to take advantage of an online request or web server [21]. It does not employ reflection, tricking, or broken packet techniques. Compared to other types of assaults, it uses the least amount of bandwidth to slow down an application or host server. When it forces the organization or request to allocate the most assets in reaction to each unit request, it is more effective.

4.1.2.2 Distributed Denial of Service

A distributed denial-of-service (DDoS) assault takes place when several systems attack a server with fake traffic, as shown in Figure 4.2. Finally, the server becomes overloaded and either crashes or stops responding to even valid requests. From 2020 to 2021, DDoS attacks increased by 341%. This was primarily because the COVID-19 pandemic forced numerous companies to convert to digital operations, which inevitably increased their vulnerability to hackers. A DDoS attack is one of the most feared cyberattacks. A well-planned DDoS attack may be very difficult to avoid and equally difficult to stop. Even the most cutting-edge IT companies' servers

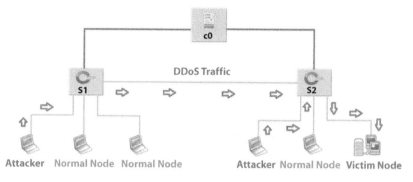

Figure 4.2 Working model - distributed denial of service.

are susceptible to them; the attacks may start at any time and the server will become unusable. In 2018, GitHub was hit by the then-largest-ever DDoS attack, which drowned their servers with over 120 million records packets each single subsequent time. The basic idea is the same regardless of the size of the onslaught: overwhelm a server with requests that it cannot process. Do this repeatedly till it clatters or stops reporting. Repairing service interruptions can frequently take hours and result in significant financial losses [20].

Application Layer Assaults
The server creates the reply to a received customer appeal at the application layer. For instance, when a person types http://www.nkr.com/resources/ into their browser, an HTTP demand for the resource page is transmitted to the server. The server collects the page-related data, compiles it into a reply, and directs it to the browsing software once more. The application layer is where the information is fetched and packaged. When a hacker uses various bots or technologies to bombard the server with requests for the same resource repeatedly, this is known as an application layer attack. One example of this is repeatedly inquiring a server to produce PDF booklets. The server cannot identify an attack because the IP address and other identifiers vary with each request [22].

Protocol Attacks
All of the volumes of web servers and other assets, such as firewalls, are used by procedure assaults. They mark the board unattainable by baring hovels in network layer and transport layer. An instance of a protocol attack is a synchronization flood, in which the invader blasts the target with a large number of handclasp needs for the TCP with copied source

IP addresses. The goal is to overwhelm, since the battered servers make an effort to answer all joining appeal, but the last handclasp never takes place [23].

4.2 Related Works

The main objective of the work is fully focused on attacks on software-defined network. A method, SDN Secure Control and Records Plane (SECOR), is a potent procedure that employs novel activates to identify and prevent Denial of Service assaults in both the control and statistics planes. Additionally, in order to replicate and study the consequences of denial of service assaults on the hardware switch and controller for SDN, The SDN network testbed was established. Additionally, they confirmed and tested SECOD's capability to recognize and prevent denial of service assaults on SDN. The examination findings demonstrate that denial of service attacks may be recognized and countered, strengthening SDN's security characteristics. They found that a dynamic threshold would increase both the controller and switch's resource efficiency and flexibility [3].

SDN-based application that integrates with the OpenDayLight controller to record and analyse traffic heading toward the victim in a time period of between 100 and 150 sec, the work identifies and mitigates DDoS assaults and restricts them at their source [24]. Using SDN standards, areas of strength, limits, and the fact that the SDN specification specifies packet forwarding to the controller, the work focuses on a solution that works exceptionally well for SDN [4].

There is a strategy to counter SIP DoS attacks. Conserving bandwidth and making full use of the restricted system cache considerably increases performance while fighting against SIP DoS assaults when compared to the conventional protection strategy. Additionally, they tested this plan and confirmed its effectiveness. The performance has greatly improved with this new plan. However, the simulator shows that the size of the low priority queue is still shrinking gradually [5].

HDB is a technique based on a sender's past event. The HDB system estimates the proportion of stream of traffic that should be reflected to intrusion and detection system. The controller consults the HDB to get the sender's incident details. Traffic flow copied to intrusion detection system is sharp by way of provided minimal traffic if the dispatcher consumes no incidents logged in HDB. Traffic flow imitated to intrusion and detection system is well-defined as the provided maximum stream of traffic if there is an incident and the incident is equal to or greater than the specified threshold. Otherwise, traffic ratio is used to determine how much traffic is mirrored to IDS. In comparison to traditional schemes, the HDB strategy decreases stream of traffic in the connection to intrusion detection system by a regular of 54.1% after the trial case [6].

Also to be considered is the effect of the denial of service assault on the bandwidth between two multitudes in the software-defined network and how it affects the OpenDaylight and Python-based software defined network controller. After initiating a denial of service assault, the results on OpenDaylight and Python-based software defined network controller displayed a low bandwidth result. A DoS attack still had a detrimental impact on a host's bandwidth even after a valid user connected to a server. This result is brought about by the switch's memory limitations, which prevent it from adding a movement slab for a valid handler once a flow timeout has occurred. Another factor contributing to the unfavorable outcome is the controller congestion experienced while handling packet-In events and attempting to connect flow tables whose buffer has been removed from a switch. In order to stop any traffic from isolating the SDN architecture, it may be possible to apply a packet rate restriction. But this needs to be done cautiously, especially if there are several hosts on a network that are simultaneously accessing the same server [7].

Regarding the viability of the Distributed Self-Organizing Map (DSOM) for DoS attack detection, in the suggested method, several DSOMs are active and individually detecting DoS assaults at various points. To eliminate the map divergence, they are combined into a single SOM in a weighted sum method [16]. The trials using actual data demonstrated that DSOM is capable of matching the original SOM approach's detection performance. They modify the DSOM to fit a Software-Defined Network (SDN) environment in future development [8].

The strategy uses two new information measures, the RE metric and the ID metric, to classify low-rate distributed denial of service assaults. The software-defined network information plane is dangerously threatened by the tiny volume of malicious traffic. Although it is quite difficult to identify this kind of assault, it is crucial to study it as soon as it manifests. The amount of control events has the most influence on the SDN controller layer [9]. The controller layer experiences congestion as the quantity of events rises, which results in a reduction in server resources. The only Shannon entropy approach available in this case is insufficient to detect the false alert. As a result, RE can be used as information distance metric for low-rate DDoS attacks [10].

To identify and counteract DDoS attacks, they created the lightweight DOCUS model. As an additional unit for the Python-based checker, they built DOCUS. The three components that make up DOCUS are changed switch performance, finding, and moderation [20]. Using the amount of synchronization and final acknowledgement packets at the organizer of the network, they modified the switch behavior; the changed switch behavior accurately counts the amount of half-open and whole networks. Because the detection module employs the stateless CUSUM method, DOCUS is both generic and stateless. They altered the CUSUM algorithm in the detection module to identify DDoS assaults and distinguish flash traffic from attack traffic [11]. Table 4.1 contains malicious attack detection methods and their limitations.

Table 4.1 Relative study of connected works.

Paper	Finding method	Factors	Benefits	Limits
12	SDN Secure Control and Data Plane (SECOD)	MAC & Desc IP	Dynamic threshold increase for both the controller and switch's resource efficiency and flexibility	Only used for single controllers
13	OpenDayLight controller (ODLC)	IP & MAC Web Url	Minimum time required to analysis the hacker in the system	Packet forwarding to specific machine leads to many type of attacks for the system
14	SIP DoS attack strategy	In time and out time of the packet	low priority queue is still shrinking gradually	No of data packets increased its automatically slow down the process
15	Hdb technique	Traffic ratio, bandwidth	Hdb technique based on a sender's past event.	Mirroring is difficult if the traffic is high

(Continued)

Table 4.1 Relative study of connected works. (*Continued*)

Paper	Finding method	Factors	Benefits	Limits
23	Effects of the DoS attack on the bandwidth	Bandwidth	Bandwidth among two hosts in the software defined network	If the bandwidth is low its automatically reduced the efficiency
24	Distributed Self-Organizing Map (DSOM)	Time, Bandwidth	Automatic detection of new types of attacks	Takes time to identify the attack in the initial stage
25	RE metric and the ID metre, to identify low-rate DDoS attacks	MAC address	Easy identification DDoS attack	It's difficult to find the attack during the heavy traffic
26	Lightweight DOCUS model	MAC Address	Less memory is needed to implement the model	Difficult for huge traffic

4.3 Proposed Work for Threat Detection and Security Analysis

The proposed work is used to detect malicious DDoS and DoS attack using following modules, traffic collection, feature selection using entropy, malicious traffic detection and traffic mitigation, as shown in Figure 4.3. Malicious nodes are detected by two stages using a machine learning approach. Under stage 1 traffic is grouped using K-Medoids and under stage 2 traffic is filtered using multinomial regressions.

4.3.1 Traffic Collection

4.3.1.1 Data Flow Monitoring and Data Collection

With the increase in usage of the internet, the flow of data across the globe is substantially increased. The development in internet technologies, cloud services, networking capacities, hardware and software, data

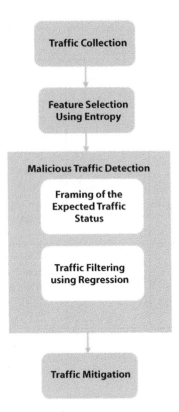

Figure 4.3 Proposed work.

flow monitoring, and data collection plays a vital role for network people, and the number of security analysts has grown rapidly in recent years. Cyberattacks are growing more complicated, clever, and targeted. Despite these complications, the essential function of data flow monitoring, data collecting, data processing, data analysis in quick time, and monitoring the security of the system remains unaltered. This plays a critical role in detecting and responding to network intrusion. Many leading companies have instead initiated the use of new categories or sorts of network data that can be collected. Security experts have been enabled by the above-said features to obtain a better understanding of their network's activities, and assess its security in a better way.

4.3.1.2 Purpose of Data Flow Monitoring and Data Collection

It is the role of network people or security experts to study the data which contains critical information. The majority of data flowing in the network is free from critical information. Only a very small percentage contain critical information. So the expert's role is to find the bits correctly, gather all affected bits, analyze them and safeguard them from the normal user. Proactively, security experts may use real-time network data monitoring, testing, and analysis to assist in uncovering network vulnerabilities, assessing performance, evaluating service levels, and even detecting suspicious activities. Regardless of the organized form of individual network data, different security risks, assaults, and intrusions can create variances in network traffic type, volume, source, and destination.

4.3.1.3 Types of Collection

Everyone who uses the internet is producing data. The nature of data produced by individuals, companies, and hosting agencies varies depending on the purpose. The purpose may be training data, data used for testing, detection of vulnerabilities, or any incidental purpose. Data collection and analysis are based on various factors such as the quality of collections, tools available for collection, and various capacities to efficiently and effectively analyze the data after collection. Because of these, network data collection and analysis are becoming a complicated task. For all of these reasons, companies and security professionals must understand the many sorts of data that may be collected and, ultimately, what the data can tell them about what is happening or has happened within the network.

Packets are very important for traffic analysis in the network. A packet carries a lot of important data. It is used in all devices from basic mobile phones to large-scale industries. Because of its usage, it is the most common way of data collection in a particular place at a particular time. Packets consist of the Packet header, Payload (Original Message), and Trailer (to show the end of the message). The packet header is responsible for the packet to be delivered to the destination address. The packet header contains information such as source addresses, destination addresses, source port, destination port, and type of protocol, etc. This information helps in packet delivery from source to destination through the network. The payload contains the original information which is present as it is or sometimes the information is encrypted.

Source Addresses (IP_{src}) and Destination Addresses (IP_{des})
For the packet transmission between two nodes, it is necessary that the packet should have the source IP addresses IP_{src} (originating) and destination IP addresses, IP_{des}. This may also be used to determine "normal" traffic

when gathered. Capturing a packet capture, for instance, might indicate to DDoS attacks or some other attack like botnets where IP_{src} for traffic occur outside of standard ranges of IP or more concentrated when we have a comparison with genuine users.

Source (Port$_{src}$) and Destination Port (Port$_{des}$)
The transmission packets are routed to various ports according to its needs and an examination of their frequency over time can create a baseline for monitoring for abnormalities. Security analysts will do more analysis and investigate when any of the following happens.

- abnormal port scans
- abnormal traffic at any of the ports

Packet Content
All generated packet has two parts, namely viz. packet header (responsible for packet delivery) and packet content (original message and trailer to denote the end of message). For security purposes, both parts may be taken for examination. For example, if we take any antivirus software, packet headers are selected for examination to see if extraordinary quantities of network activity are targeting known susceptible programmers or including strange source IP information. Similarly, packet content of the same may be examined to see whether malicious code is there or whether odd application commands are present, which can indicate foreign cyber-attacks. However, the transmission protocol of the packet, as well as the types and placement of security and monitoring devices utilized, might limit packet content examination.

Flow-Level Data
Considering present internet speed and data transfer, one has to think beyond collecting and analyzing packet-level data. As a result, the practice of collecting flow-level data has grown, offering a macro-level perspective of network activity by analyzing groups of packets that have similar destinations, sources, protocol types, or other information indicated in the packet's header. Flow analysis can help monitor network performance, application health, or host activity by analyzing similar packets together, as well as identifying unexpected network traffic that may indicate a potential intrusion. When adopting flow-level data collection, businesses must decide where the data will be collected as well as the extent of that data collection. First, while flow data collecting can occur at any point inside a

network and even at several network points at the same time, it is generally most successful when linked with network edge nodes that monitor data going in and out of a local area network. At the same time, flow data may be acquired using either a "depth-first" or a "breadth-first" strategy. The former narrows the kind of flow data gathered to fit certain criteria found in packet headers, whereas the latter aims to collect as much information as possible in order to gain a comprehensive picture of overall network activity.

Connection-Based Data
As of now, we have discussed flow-based data collection and packet-based data collection. Both will be like a black box which provides comprehensive network information for review. But connection-based data on the other hand, like a white box, provides a deeper level of understanding about the traffic flow in an environment. It aggregates the traffic of the network between any two parties.

4.3.2 Feature Selection Using Entropy

Initially the abnormal status of the traffic flow is detected using Rényi's quadratic entropy. The Rényi entropy is named after Alfréd Rényi, who looked for the most general way to quantify information while preserving additivity for independent events. Figure 4.4 contains Port Address (Src_p, Dst_p), IP address (Src_{IP}, Dst_{Ip}), Physical address (Src_{MAC}, Dst_{MAC}), VLAN ID

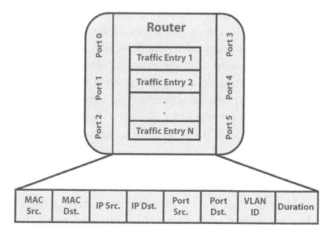

Figure 4.4 Feature selection.

and Duration are given input to Rényi's quadratic entropy. Identical nodes are detected using Rényi's quadratic entropy using equation 4.1. The Rényi entropy is significant within the field of subatomic particles because it may be utilized to quantify connectivity. The Rényi entropy being a factor may be computed directly with in Heisenberg XY turn chain system since this comprises an automorphic functional in regard to a certain specific group of the modularity group. The broadest approach to data quantification maintains properties for unrelated occurrences. The Rényi entropy gradually ranks all nonzero likelihood occurrences same as tends to zero, irrespective of actual likelihood. The Rényi entropy is simply the logarithmic of the amount of the assistance of X in the bound for 0. When $0 < \alpha < \infty$ and $\alpha \neq 1$. The Rényi divergence the value is $\alpha = 0, 1, \infty$ by captivating a maximum, and in specific the maximum $\alpha \to 1$.

$$H(B) = \begin{cases} -\log \sum_{x=1}^{n} P_x, & \text{where } P_x = b_x/n \\ \text{mimimal} & \text{for concentrated samples} \\ 0 & \text{for identical samples} \end{cases} \quad (4.1)$$

4.3.3 Malicious Traffic Detection

The internet is all about the movement of data, or traffic. Everyone is connected with the internet and when internet is connected, there is a flow of data between sender and receiver. We cannot predict whether all incoming data is good. There are some chances that through wanted data, malicious stuff is prevented from accessing our personal system. It is the purpose of your antivirus solution's harmful traffic detection capability to keep your computer safe. If we were to rank the different elements of your endpoint security in order of priority, detecting malicious traffic would undoubtedly come top. A malicious traffic detection system continually analyses traffic for any indications of suspicious links, files, or connections that are made or received. Advanced harmful traffic detection skills can determine whether a suspicious link is a type of malicious traffic originating from bad URLs or C2 sites in order to identify malicious traffic. Typically, it compares the link to a massive quantity of security data collected from hundreds of millions of devices around the world. This safeguards against both known and unknown dangers. When malicious HTTP requests reach the command and control servers, they send a message to your hacked PC or Mac, enlisting it in their bigger zombie army known as botnets. This communication may be as easy as keeping a timed beacon on your PC so that hackers who have hijacked your PC can keep track of how many such PCs are in their inventory (yes, they have an inventory!). Alternatively, attackers can give orders to initiate harmful acts such as data theft or a ransomware attack. The malware must access your system in order for a command and control assault to take place. This is most commonly accomplished through phishing emails and social engineering assaults.

4.3.3.1 Framing of the Expected Traffic Status

Kaufman and Rousseeuw presented the K-Medoids (also known as Partitioning around Medoid) method in 1987. A medoid is defined as the point in the cluster having the fewest dissimilarities to the other points in the cluster. E = |Pi - Ci| is used to calculate the dissimilarity of the medoid (Ci) and object (Pi) are given in equation 4.2.

$$c = \sum_{Ci}^{n} \sum_{Pi \varepsilon Ci}^{n} |Pi - Ci| \qquad (4.2)$$

4.3.3.2 Traffic Filtering Using Regression

The multinomial regressions are used to segregate the traffic are given in equations 4.3 & 4.4.

$$Prob_{NT} = 1 - \sum_{M'=1}^{m} prob(Z_x = 0|Y) e^{Dep fun(Y)} \qquad (4.3)$$

$$Prob_{AT} = e^{Dep fun(Y)} \Big/ 1 + \sum_{M'=1}^{m} e^{Dep fun(Y)} \qquad (4.4)$$

Normal traffic and malicious traffics are detected using the equations 4.3 and 4.4.

4.3.4 Traffic Mitigation

The first step is to select a medoid. It is selected by choosing k random points from the given pool of n data points. We can use any of the conventional distance metric methods which will connect each given point to the nearest selected medoid. Do the following as the cost decreases. For each of the medoid point a and non-medoid point b, swap a and b, then assign each point the medoid which is nearer to it. Once it is done recalculate the cost. If the recalculated cost seems to be higher than the previous cost, undo the last step or else repeat the process.

4.4 Implementation and Results

Experiments for configuring SDN are implemented on mininet SDN testbed. Mininet is a SDN emulator. Synthetic traffics are generated using hping3 command line utility. Mininet are configured on a Dell Inspiron with 16 GB RAM, an Intel Core i7-1011 CPU and a 64-bit OS, running Ubuntu PC. Syntactic traffic are mitigated using the POX controller. POX Controller will collect the traffic on every 30s from all the OVS (OpenFlow vSwitches) associated with SDN environment. POX controller will implement new mitigation rules to deny the malicious traffic entering into home networks. Table 4.2 shows dataset details; it contains synthetic and KDD dataset. Trace/packet types and its percentage are given in Table 4.2. Several evaluation parameters were used to evaluate the classification efficiency of the proposed system: Accuracy (AC), Detection Rate (DR), Precision (PR), True Negative Rate (TNR) and False Alarm Rate (FAR). AC: Based of all the observations in the study tests, it counts the amount of instances the model properly detected. The model considers both TP and TN while determining its accuracy given in equation 4.5.

$$AC = \frac{(TP+TN)}{(TP+FP+TN+FN)} \quad (4.5)$$

PR: By partitioning the entire amount of categorized assault inspection, the model represents the amount of recognized assaults and the insights of those assaults that were found given in equation 4.6.

$$PR = \frac{TP}{(TP+FP)} \quad (4.6)$$

FAR: To correlate with regular insights designated as an assault, the entire amount of regular facts in the dataset is halved by the entire amount of regular facts given in equation 4.7.

Table 4.2 Dataset description.

Details	Synthetic dataset	KDD dataset [29]
Trace Type and Traffic Percentage	TCP & 60%	TCP & 78%
	UDP & 32%	UDP & 12%
	ARP & 5%	ARP & 3%
	ICMP & 3%	ICMP & 7%
Time Window (Traffic Request)	30 seconds	-
Trace Size	1024 bytes	80 - 1024 bytes
Number of Traces	1278225 Trace	1068800 Trace

$$FAR = \frac{FP}{(FP+TN)} \qquad (4.7)$$

TNR: It details the proportion of actual regular cases that the recognition method predicts to be regular given in equation 4.8.

Figure 4.5 shows before performing mitigation services, traffic contains malicious traffic and normal traffic. Packet retransmission occur due to malicious traffic. Figure 4.6 shows malicious traffic is reduced after implementing the mitigation process. In Figures 4.5 & 4.6, normal traffic is projected in the gray area and malicious traffic is projected in the orange area.

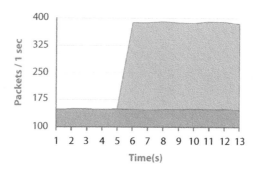

Figure 4.5 Before mitigation.

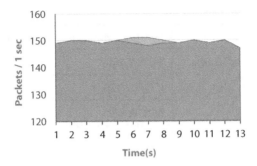

Figure 4.6 After mitigation.

$$TNR = \frac{TN}{(TN + FP)} \qquad (4.8)$$

Figure 4.7 shows normal traffic time bins and proposed system perform better than AD-NB and AD-MLP on normal traffic processing. Figure 4.8 shows proposed systems better time bins on implementing mitigations process for different volume of data and perform better than AD-NB and AD-MLP.

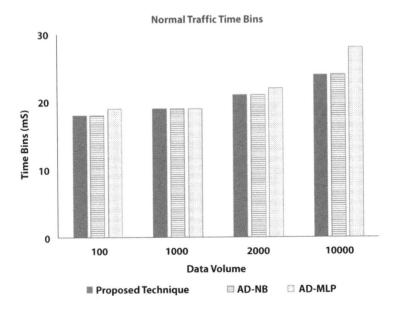

Figure 4.7 Time bins for normal traffic processing.

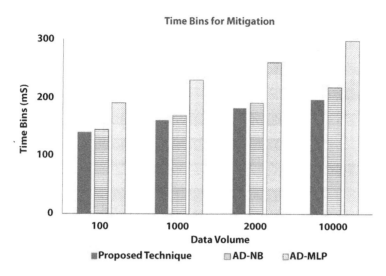

Figure 4.8 Time bins for mitigation process.

Figure 4.9 shows CPU usage on malicious attack detection for different volume of data. Proposed system uses POX controller-based attack detection. POX controller will perform mitigation services based on dynamic access control list.

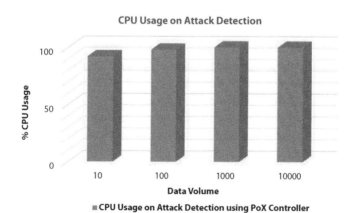

Figure 4.9 CPU usage on attack detection.

Table 4.3 Performance of attack detection for synthetic data set.

Detection technique	Attack type	Accuracy (%)	Precision (%)	Recall (%)	F-measure (%)	FP rate (%)
Proposed Technique	DDoS	91	91	78.31	86.19	2.82
	DoS	92	92	79.1	85.9	2.9
AD-NB	DDoS	84	86	58.2	69.8	10.4
	DoS	83	85	58.5	70.1	11.4
AD-MLP	DDoS	68	65	38.2	49.8	14.2
	DoS	67	64	39.2	48.8	15.5

Table 4.4 Performance of attack detection for KDD data set.

Detection technique	Attack type	Accuracy (%)	Precision (%)	Recall (%)	F-measure (%)	FP rate (%)
Proposed Technique	DDoS	90	90	79.1	87.19	3.2
	DoS	91	91	78.1	84.9	3.9
AD-NB	DDoS	85	85	59.2	70.2	11.2
	DoS	84	84	60.6	69.3	10.3
AD-MLP	DDoS	70	69	35.2	49.1	15.1
	DoS	69	68	38.2	49.2	16.2

Table 4.3 shows proposed techniques perform better than AD-NB and AD-MLP in terms of attack detection, results accuracy, precision and FP rate. Table 4.3 shows malicious attacks detection for simulated dataset. Table 4.4 shows proposed techniques produced good results in terms of attack detection, accuracy and FP rate compared to AD-NB and AD-MLP for KDD data set.

4.5 Conclusion

In this research proposed techniques are used to detect the DoS and DDoS traffic and perform mitigation services against malicious flood. Proposed systems are implemented on mininet SDN emulator and synthetic traffics are generated using hping3 command line utility. POX controller uses dynamic access control list for implementing mitigation process. Proposed techniques have high accuracy on attack detection and less FP rate compared to AD-NB and AD-MLP techniques. In future work, other types of malicious assault detection and network load-sharing conceptions are implemented using POX controller.

References

1. Anju Markose, Shebin Sharief, J Ramprasath, N Krishnaraj, Survey on Application of IoT and its Automation, *International Journal of Advanced Engineering Research and Science*, Vol. 8, p. 6, 2021.
2. Balasamy K, Krishnaraj N, Ramprasath J, Ramprakash P, A secure framework for protecting clinical data in medical IoT environment, *Smart Healthcare System Design: Security and Privacy Aspects*, 2021. https://doi.org/10.1002/9781119792253.ch9
3. Balasamy K, Suganyadevi S, A fuzzy based ROI selection for encryption and watermarking in medical image using DWT and SVD, *Multimed Tools Appl* 80, 7167–7186, 2021.
4. Dr M Balakrishnan, Dr AB Christopher, Dr AS Murugavel, J Ramprasath, Prediction Of Data Analysis Using Machine Learning Techniques, *Int. J. of Aquatic Science*, Vol. 12, Issue 3, pp. 2755-2762, 2021.
5. J Ramprasath, AB Arockia Christopher, M Balakrishnan, AS Muthanantha Murugavel, Denial of Service Malevolent Traffic Identification and Prevention in Software Defined Networking, *2nd International Conference on Advance Computing and Innovative Technologies in Engineering (ICACITE)*, pp. 400-405, IEEE.
6. J Ramprasath, Dr S Ramakrishnan, P Saravana Perumal, M Sivaprakasam, U Manokaran Vishnuraj. Secure Network Implementation using VLAN and ACL, *International Journal of Advanced Engineering Research and Science*, Vol. 3, Issue 1, pp. 2349-6495, 2016.
7. J Ramprasath, M Aswin Yegappan, Dinesh Ravi, N Balakrishnan and S Kaarthi, Assigning Static Ip Using DHCP In Accordance with MAC, *International Journal for Trends in Engineering & Technology*, Vol. 20, Issue 1, 2017.
8. J Ramprasath, P Ramya, T Rathnapriya, Malicious attack detection in software defined networking using machine learning approach, *International Journal of Advances in Engineering and Emerging Technology*, 11 (1), pp. 22-27, 2020.

9. J Ramprasath, V Seethalakshmi, Mitigation of Malicious Flooding in Software Defined Networks Using Dynamic Access Control List, *Wireless Personal Communications* (2021). https://doi.org/10.1007/s11277-021-08626-6
10. K Balasamy, N Krishnaraj, K Vijayalakshmi (2022) Improving the security of medical image through neuro-fuzzy based ROI selection for reliable transmission https://link.springer.com/article/10.1007/s11042-022-12367-4
11. K. Balasamy, D. Shamia. Feature Extraction-based Medical Image Watermarking Using Fuzzy-based Median Filter, *IETE Journal of Research*, 2021.
12. M Sakthivadivel, N Krishnaraj, P Ramprakash, Utilization of big data in oil and gas industries using Hadoop MapReduce technology and HiveQL, *Global Journal of Multidisciplinary and Applied Sciences* 1 (2), 52-57, 2013.
13. M. J. Abinash, Sountharrajan, S., Bhuvaneswari, R., & Geetha, K. (2022). Identification and Diagnosis of Breast Cancer using a Composite Machine Learning Techniques. *Journal of Pharmaceutical Negative Results*, 13(4), 78-85.
14. N Krishnaraj, RB Kumar, D Rajeshwar, TS Kumar. Implementation of Energy Aware Modified Distance Vector Routing Protocol for Energy Efficiency in Wireless Sensor Networks, *International Conference on Inventive Computation Technologies*, 2020.
15. N Krishnaraj, S Smys. A multihoming ACO-MDV routing for maximum power efficiency in an IoT environment. *Wireless Personal Communications* 109 (1), 243-256, 2019.
16. P Jayasri, A Atchaya, M Sanfeeya Parveen, J Ramprasath, Intrusion Detection System in Software Defined Networks using Machine Learning Approach, *International Journal of Advanced Engineering Research and Science*, Vol. 8, Issue 8, 2021.
17. P Ramprakash, M Sakthivadivel, N Krishnaraj, J Ramprasath.: Host-based Intrusion Detection System using Sequence of System Calls, *International Journal of Engineering and Management Research*, Vandana Publications, Vol. 4, Issue 2, pp. 241-247, 2014.
18. Prakhar, K., Sountharrajan, S., Suganya, E., Karthiga, M., & Kumar, S. (2022, April). Effective Stock Price Prediction using Time Series Forecasting. In *2022 6th International Conference on Trends in Electronics and Informatics (ICOEI)* (pp. 1636-1640). IEEE.
19. Ramprasath J, Seethalakshmi V, Improved Network Monitoring Using Software-Defined Networking for DDoS Detection and Mitigation Evaluation, *Wireless Personal Communications*, 116, 2743–2757, 2021.
20. Ramprasath J, Seethalakshmi V, Secure access of resources in software-defined networks using dynamic access control list, *International Journal of Communication Systems*, Vol. 34, no. 1, pp. 1-12, e4607, 2020.
21. Smys, S., B. Abul, and W. Haoxiang. Hybrid Intrusion Detection System for Internet of Things (IoT), *Journal of ISMAC* 2, no. 04, 190-199, 2020.

22. Suganyadevi S, Shamia D, Balasamy K (2021) An IoT-based diet monitoring healthcare system for women. *Smart Healthc Syst Des Secur Priv Asp.* https://doi.org/10.1002/9781119792253.ch8
23. V Ponmanikandan, J Ramprasath, KS Rakunanthan, M Santhosh Kumar, An ecosystem for vulnerable traffic analysis and mitigation services in software defined networking, *International Research Journal of Engineering and Technology*, Vol. 7, Issue 6, pp. 5287-5295, 2020.
24. Yahui Li, Zhiliang Wang, Jiangyuan Yao, Xia Yin, Xingang Shi, Jianping Wu, Han Zhang, MSAID: Automated detection of interference in multiple SDN applications, *Elsevier Journal of Computer Networks*, 153 (2019) 49–62, https://doi.org/10.1016/j.comnet.2019.01.042
25. H. Anandakumar, R. Arulmurugan, and C. C. Onn, Big Data Analytics for Sustainable Computing, *Mobile Networks and Applications*, Vol. 24, no. 6, pp. 1751–1754, Oct. 2019.
26. H. Anandakumar and R. Arulmurugan, Next Generation Wireless Communication Challenges and Issues, *2019 Third International conference on I-SMAC (IoT in Social, Mobile, Analytics and Cloud) (I-SMAC)*, Palladam, India, 2019, pp. 270-274.
27. A. Haldorai, A. Ramu, and M. Suriya, Enterprise Architecture for IoT: Challenges and Business Trends, in *Business Intelligence for Enterprise Internet of Things*, pp. 123-138, 2020. doi:10.1007/978-3-030-44407-5_6
28. H. Anandakumar, R. Arulmurugan, The Impact of Big Data Analytics and Challenges to Cyber Security, *Handbook of Research on Network Forensics and Analysis Techniques*, IGI Global, Publications. Chapter 16, ISBN13: 9781522541004, ISBN10: 1522541004, EISBN13: 9781522541011.
29. http://kdd.ics.uci.edu/databases/kddcup99/kddcup99.html

5

Privacy Enhancement for Wireless Sensor Networks and the Internet of Things Based on Cryptological Techniques

Karthiga, M.[1*], Indirani, A.[2], Sankarananth, S.[3], S. S. Sountharrajan[4] and E. Suganya[5]

[1]*Department of Computer Science and Engineering, Bannari Amman Institute of Technology, Sathyamangalam, Tamilnadu, India*
[2]*Department of Artificial Intelligence and Machine Learning, Bannari Amman Institute of Technology, Sathyamangalam, Tamilnadu, India*
[3]*Department of Electrical and Electronics Engineering, Excel College of Engineering and Technology, Namakkal, Tamilnadu, India*
[4]*Amrita School of Computing, Amrita Vishwa Vidyapeetham, Chennai, Tamilnadu, India*
[5]*Department of Information Technology, Sri Sivasubramaniya Nadar College of Engineering, Kalavakkam, Chennai, Tamilnadu, India*

Abstract

Nearly everyone has at least one internet-connected gadget in the connected world these days. As the number of these gadgets grows, it's critical to create a security policy to reduce the risk of abuse. Negative actors may employ internet-connected gadgets to gather private information, hijack identities, jeopardize banking details, and covertly monitor—or observe. The setup and use of the gadgets can assist in stopping this kind of behavior if a few safeguards are taken. Although Wireless Sensor Networks (WSNs) have a wide number of uses, the safety of these WSNs is becoming increasingly crucial as sensor nodes get more intricate. Due to the placement of sensors in isolated places and the geographic dissemination of WSNs, significant security risks exist. Researchers are now looking into potential solutions in this new field as a result. A crucial component of WSN security is discussed and summarized. Numerous encryption methods, including symmetric keys and public keys, are investigated. These methods include identity-based

---

*Corresponding author: mkarthiga22@gmail.com

S. Sountharrajan, R. Maheswar, Geetanjali Rathee, and M. Akila (eds.) *Wireless Communication for Cybersecurity*, (105–128) © 2023 Scrivener Publishing LLC

cryptography (IBC), pair-based encryption, and elliptic curve cryptology. This article proposes a data protection method which is an enhanced version of the proposed Diffie-Hellman algorithm that requires reduced computation and response time. By creating a hashing of each value that is sent over the network, the Diffie-Hellman has been changed to make it more secure against assaults. The proposed method's security against different assaults has been examined. It has also been examined in terms of the length of time required for encryption and decryption, computation, and key creation for data of various quantities. The proposed strategy outperforms the existing ones in most instances, according to a comparison with them.

Keywords: Wireless sensor networks, cryptography, encryption, decryption, pseudo random number generator, LEACH, Diffie Hellman algorithm

5.1 Introduction

The WSN is a flexible network with many kinds of sensor nodes based on the networking mechanism used: sinking node, ground station (GS), sensor nodes, and clustering heads. The data is delivered through the sensors to the clusters, which further send it to the GS via a modern communication procedure. Sensors are utilized to detect and send data in a variety of settings. Sensors in various real-time applications perform unique tasks such as neighbor node detection, intelligent sensing, confidentiality objective checking, tracing, node location, synchronization, and effective routing across nodes and GSs.

Secured data transfer is a big concern in WSN as there are numerous adversaries on the connection who can assault or counterfeit the data. This WSN data protection study focuses on studies published between 2016 and 2022. Sensor networks have constrained computing, memory, and power capabilities [Akyildiz *et al.* (2002)]. There have been a substantial number of studies published on digital security in WSNs. Elliptic Curve cryptography (ECC) is employed in several suggested research methodologies [Elhoseny *et al.* (2016) and Mahmood *et al.* (2019)]. Ullah *et al.* (2018) utilize it in conjunction with Advanced Encryption Standard (AES), in which the key is created via ECC. AES, on the other hand, is a block encryption technique with a long processing time based on the length of the keys and message.

To provide secure data exchange, ECC is frequently integrated with deoxyribonucleic acid (DNA) [Tiwari *et al.* (2018)]. A chaotic mapping is a super secure quantitative approach, and Rivest-Shamir-Adleman (RSA) is widely employed in WSNs to protect data. [Wang *et al.* (2018),

Ahmad *et al.* (2018)] Because the WSN is a complex ecosystem, a few recommended research methodologies, such as AES hybrid Elliptic Curve Cryptography (HECC) [Tiware *et al.* (2018)] and Elliptic Curve cryptography Genetic algorithm (EGASON) [Mahmood *et al.* (2019)], are thought to be the most suited. This study offered a novel cryptographic technique to secure data in wireless sensor networks. For data forwarding, the lower-energy adaptable clustering hierarchy (LEACH) algorithm is utilized. For data encryption, the suggested research method uses Hybrid Diffie-Hellman. Because Diffie-Hellman is indeed an asymmetrical technique, it provides a high level of security. Such an encryption method requires lower responsiveness and time complexity.

The rest of the article is aligned as follows. Section 5.2 reveals the representation of the setting of the WSNs. Section 5.3 reveals the various researches held with respect to the proposed approach. Section 5.4 details the proposed algorithmic technique employed to secure the data transmitted in the WSNs and IoT networks, and section 5.5 illustrates the experimental results.

5.2 System Architecture

The WSN is made up of several sensing nodes that interact with one another. A sink unit and a GS help transmit the data between sensors in the connection between two nodes. The main idea behind WSNs is to remotely transport data among nodes. Figure 5.1 depicts a high-level representation of WSNs, including sink nodes, sensors, and GS.

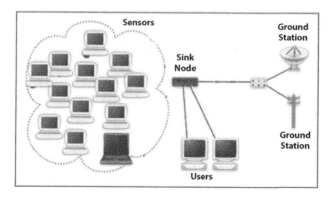

Figure 5.1 System architecture.

5.3 Literature Review

WSN input security is a key concern. The most important difficulty is to protect the information using an approach with a short calculation time, quick reaction time, minimal energy usage, and restricted bandwidth. A number of algorithms are proposed to protect knowledge in order to overcome obstacles such as power shortages, coverage concerns, and the creation of limited use of data measures. This part is further subdivided into the following sections: an elliptic curve, effective block ciphers, AES, chaotic-maps, RSA, and several additional algorithms.

Elhoseny *et al.* (2016) developed a new approach known as EGASON. This method employs ECC for generating keys as discussed by [Mahmood *et al.* (2019)] and then employs the evolutionary algorithm methodology for encrypting data and deciphering. In contrast to symmetrical key processes, ECC may be vulnerable to brute-force attacks and have a large computational cost [Chaudhry *et al.* (2020), Mansoor *et al.* (2019)]. The approach in Somsuk *et al.* (2018) has the potential to break the usage of pseudo-random numeric generators (PRNGs). Santhosh *et al.* (2017) demonstrate strong encryption. On considerations of security, such a method solely employs the exclusive OR (XOR) function for encrypting, deeming it vulnerable to recognized and selected assaults [Daemen *et al.* (1991)].

A further study introduced elliptic curve cryptography-key managing (E-KM) [Singh *et al.* (2017)], which includes a key structuring mechanism. This study approach is vulnerable to collision screening attacks [Wiener *et al.* (1998)]. The most recent research confirms that implementing ECC puts it in danger of erroneous curve assaults [Neves *et al.* (2017)]. For safe transmission in WSNs as discussed by Viswanathan *et al.* (2019) employ elliptic curve key cryptography utilizing beta as well as gamma factors. A number of approaches employ ECC for WSNs, resulting in a demand for a new safe encryption technique for WSNs. As a novel key extraction strategy, Ghani *et al.* (2019) employed symmetrical key cryptography for device authentication within WSNs as well as the development of shareable keys for both encrypting and decrypting.

Ullah *et al.* (2018) presented HECC as a novel key generation technique in their investigation. This secret key is capable of encryption/decryption as well as node authentication. The usage of a randomized number maker in this study can be fractured using Reeds' approach [Reeds (1977)]. As AES is utilized to guarantee data protection, it is also vulnerable to biclique threats [Bogdznav *et al.* (2011)].

The study of Tiwari et al. (2018) employs ECC and DNA for information cryptography and deciphering. To allocate genes, DNA is thrown off, and such genes are employed for encrypting data. This research is safe from timing and Simple Powers Analysis (SPA) assaults. This approach, however, is vulnerable to Man-in-the-Middle (MIM) attacks. Suresh et al. (2017) conducted a study on the IoT ecosystem. Double cryptographic architecture-secure network connection (D-SN) is an information security technology that encrypts information with DNA gene sequences using RSA and Data Encryption Standard (DES). The usage of a small open and secret key length will result in a security concern. Nitaj et al. (2014) and Weger et al. (2002) employed RSA, which is separated into four steps: distribution of keys and generation, encoding, en route screening, and routing. The use of RSA in this technique renders it vulnerable to Wiener and Boneh-Durfee assaults [Roy et al. (2018)].

MRSA (Modified Rivest Shamir Adleman) is an extra encryption method that enhances the current RSA [Manger et al. (2001)]. This study approach is vulnerable to a chosen-ciphertext assault. Furthermore, energy usage rises as a result of triple primes. Babu et al.'s novel study (2019) uses the elliptic-curve Diffie-Hellman Key Extraction (ECDH-KE) process to provide end-to-end secrecy while also ensuring authentication and synchronization. However, this technique is vulnerable to a chosen-ciphertext assault due to the utilization of RSA [Lindell et al. (2014)]. Because of ECDH, it is subject to MIM assault [Haakegaard et al.].

Elliptic Curve Cryptography-Advanced Encryption Standards (E-AES) is another research method that utilizes ECC to produce keys for encrypting and decoding. This process is extremely secure, but it is sophisticated and entails a significant communication cost [Ullah et al. (2018)]. Another study in Li et al. (2017) delivers an improved variant of AES. AES, on the other hand, is vulnerable to biclique assaults. The secret key in this methodology is vulnerable to related-key assault. Another research method that employs AES is Advance Encryption Standard-Quadrature Phase Shift Keying (AES-QPSK), which includes or does not include a low-density parity-check (LDPC) [Khan et al. (2017)]. There is a chance that this method is vulnerable to biclique assaults. This research approach is likewise vulnerable to related-key attacks. Another study by Vangala et al. (2017) employs AES for data encryption with a hybrid mutation approach. This operation is dangerous for various attacks [Albassal et al. (2003)].

Extensive review in Vangala et al. (2018) employs AES alteration to generate factors and the first seed. This strategy is extremely concerned with biclique assaults. The [20]Pseudorandom number generator (PRNG) algorithm is vulnerable to straight cryptanalytic assault, input-based assault, backtracking assault, and other attacks. The approach in Reeds may be used to break down PRNG. The blended chaotic cryptanalysis with piecewise linear chaotic map (HTPW) is a novel research approach for key management that makes use of skew and mappings [Al-Mashhadi et al. (2015)]. Because only the XOR algorithm is employed for protection, this solution is vulnerable to selected and recognized attacks. An additional way for safeguarding data in WSNs is by employing enhanced sensitive data utilizing chaotic-based encryption (EDCB). The keys in this approach are generated using chaotic maps. Chaotic systems may indeed generate unique values; however, the usage of chaotic systems renders this approach vulnerable to ciphertext alone, known plaintext, selected clear text, and picked ciphertext.

The work in [Nidarsh et al. (2018)] employs the LEACH technique for packet transmission and then employs chaotic maps for information encrypting and decoding. In Sobhy et al. (2001), chaotic maps may be broken down using a variety of approaches. Another study by Wang et al. (2018) used logistical and Kent chaotic maps. There are a few downsides to this system, including weak passwords, constant chaotic sub-matrices, and plain-image callousness. A light block cipher (QTL) is a super-lightweight block cipher technique proposed by Li et al. (2016). This technique, however, is vulnerable to differential and polynomial attacks. Patil et al. (2017) introduce an alternative ultra-lightweight secure hash technique, linear cryptanalysis (LC). This method is vulnerable to known plaintext and selected plaintext assaults.

Maity et al. (2017) offer a novel lightweight method named lightweight pseudo-random number generator (LGA). It is feasible to conduct a known assault on this method using permutations. Another study, Solomon et al. (2018), proposes ciphertext-policy attribute-based (C-AB) cryptography as a lightweight encoding and authenticating code synthesis approach. The encryption method using this technique protects data from eavesdropping attacks. Differential attacks may have an impact on the Secure Hash Algorithm (SHA)-3. The decision-supporting system (DSS) is susceptible to timing attacks. Kocher et al. (1996) and Praveena (2017) introduces the Ultra-Encryption Standard Version 4 (UES-4) technique. UES-4 is the result of the combination of many existing cryptographic methods. Bitwise reorganization is done on the raw text, followed by bitwise tabular transit for unreadable material.

Meanwhile, several encryptions are done; the text gets extremely difficult to guess, making UES-4 resistant to brute-force attacks. This approach is susceptible to ciphertext-alone attacks (COA). The low intricacy secure algorithm (LSA) is a unique data protection study in [Li et al. (2007)].

The initial stage is to track the development of the networks. XOR operation is used to encrypt the plaintext and the key. Because nodes are merely XORed, it requires less computing time. XOR's resistance against brute-force attacks is poor.

This study by Ananda Krishna *et al.* (2018) introduces Modified Rotation XOR (MR-XOR), which is a revised variant of basic XOR. The XOR procedure is susceptible to brute-force assault. For key creation, Praveena *et al.* (2016) presented Modern Encryption Standard-Version 4 (MES-IV). This approach provides excellent defence against known plaintext, brute-force, and differential assaults. Because the ciphertext may be obtained, this research methodology is vulnerable to side-channel assaults. For data protection, this study offers an electronic signature utilizing key-management (DK) [Kumar P S *et al.* (2022)]. The asymmetrical technique increases the computing time. Gracy *et al.* (2018) work employs honey encryption (HoneyE) to generate misunderstanding. A key retrieval assault against this encryption is also possible.

Yue *et al.* (2019) adopted a mixed method in their investigation. This technique encrypts plaintext chunks using the Advanced Encryption Standard (AES) and the Elliptic Curve Encryption (ECC) technique, then compresses them using compression techniques to produce ciphertext chunks. Following that, it links the Port number and AES key encoded by ECC to generate a full ciphertext document. The writer of this study [Sountharrajan *et al.* (2020)] states that by utilizing this approach, he was able to cut encrypting time and boost privacy. Despite the fact that encryption is decreasing, utilizing AES and ECC for the sensors might limit the sensor's lifespan, resulting in a deceased sensing node and connectivity issues [Karthiga *et al.* (2021)].

Ullah *et al.* (2018) study employs AES for encrypting data, and secret keys are created via HECC. The offered technique is guarded against both front and reverse secrecy; nevertheless, this technique may be compromised by utilizing a Random Number Generator. Because AES is used, this method is subject to biclique assaults. Another technique in Manger alters the RSA by using 3 primes instead of 2, making brute-force attacks more difficult [Kumar *et al.* (2018)]. Furthermore, the usage of RSA renders this technique vulnerable to a picked-ciphertext assault.

Information safety in WSNs is a big challenge, with insufficient privacy resulting in related-key, raw text, or MIM assaults [Sountharrajan *et al.* (2017)]. A raw text assault and a MIM assault might both impact EGASON. Another significant difficulty with data protection in WSNs is that present encrypting data techniques have extremely long response times or really long calculation times [Shree *et al.* (2017)]. AES-HECC

[Lasry et al. (2016)] has a long calculation time, while EGASON [Prusty et al. (2012)] has a long reaction time. A novel safe method that consumes less computing time and responds faster is proposed. Using the information reviewed previously, a novel technique for encrypting data is offered.

5.4 Proposed Methodology

This part explains the design of the system and recommended strategies. An Advanced Secured Effective Encryption Algorithm (ASEEA) is introduced. ASEEA generates keys using a customized Diffie-Hellman (CDH) technique, which is subsequently utilized in ASEEA. The input is encoded with ASEEA and afterward with CDH finally routed using LEACH. All of these techniques are addressed in depth in the subsequent sections.

The proposed strategy is broken into three stages. Each step is intended to take as little processing and reaction time as possible. The three intended stages are listed below.

Stage 1. Effective Hashing of the input
Stage 2. ASEEA and CDH
Stage 3. LEACH.

LEACH in WSN applications is presented to allow two users with a secured data connection. LEACH provides communication security amongst sensors and safeguards against a variety of assaults. ASEEA is a novel and economical WSN cryptographic algorithm. This approach is intended to have a low computing and reaction time. Because the suggested technique doesn't really necessitate any additional difficult stages, it has the benefit of lowering processing time. The suggested technique is separated into two stages: the first involves the generation of the keys, and the next lies in applying the proposed cryptographic method.

Generating Secret Keys
During generating the secret key, the concerned parties seeking to establish communication agree on constant prime value numbers like 'j' and 'k'. The numbers are then employed in the following equation: K mod J. After the concerned parties have determined J and K, the above-said equation is utilized to construct a newer value, which is then handed to both the partners for assisting them in generating their own encryption key by utilizing their confidential values. The operation of the original Diffie and Hellman technique is detailed in Diffie et al. (1976). Because Diffie-Hellman is very

susceptible to MIM assault, an enhanced, highly secure customized variant of Diffie-Hellman called CDH is launched.

CDH (Customized Diffie-Hellman)

MIM attacks are increasingly widespread via the internet, because data may be quickly modified. CDH aids in the prevention of MIM attacks in Diffie-Hellman. The produced key by employing any cryptographic means must be safe or else the entire algorithm's safety would be compromised. Since the earliest period, hashing has been utilized to enable the encrypted connection between communicative parties. The main idea behind hashing is to turn the source data into a representation that is nonsensical and challenging to decipher. Only parties involved may defeat the hash codes, allowing hashing to function as a validation scheme. The hash code needs to be computed using the correct technique and parameters, and only the parties involved are capable of doing so.

To counter a MIM assault, the CDH technique improves Diffie-Hellman by applying hashing function. With the mastery in determining the correct functions and appropriate parameters, the recipient is the sole individual who really can compute the accurate hashes. An effective hashing method is proposed to determine the hash. All Diffie-Hellman parameters, like P_1, P_2, J, and K, are unsafe. A hash code is computed before these values are sent across the network. The values are first transferred to binary format. Following this translation, the number of 1's is determined and placed in a temporary object like Tem. Next, the mod of Tem and the values transferred to binary format is computed. The result for the value for mod is appended to the binary source input J_1. The count of zeros is determined to distinguish the hash code from actual J_1 value. Zero's count help in determining the addition of the number of special characters among the source J_1 and hash code. This modest hash value estimate will aid in protecting Diffie-Hellman weaknesses while keeping processing time to a minimum. The suggested technique will aid in the prevention of MIM attacks. The stages that follow explain how the CDH technique works.

1. The two common party's 'M' and 'N' have to decide on the parametric values like J and K
2. J and K values are communicated among each other by utilizing the hash codes. The steps for the same are as follows:
   a. First transmit the J and K values to binary form
   b. The proportion of **1's** in the output should be calculated

c. The resultant count value is stored in the variable named '**d**'. Now the mod is calculated among '**J**' and '**d**'. Again, calculate the proportion of 1's.
   d. For the count values, add the special characters
   e. Initial binary values of **J** are added with the special characters, finally the values obtained from step d is accumulated.
   f. The same type of steps is used for **K** also
3. This value is communicated to the party **N**.
4. Party **N** will compute the hash codes for **J** and **K** by utilizing the same steps from a to f.
5. Party M computes $J_1 = G^M \bmod J$
6. Then J_1's hash is computed using the steps from a to f.
7. Party N computes $K_1 = G^N \bmod K$
8. Then **K**'s hash is computed using the steps from a to f.
9. Both resultant values are communicated by the parties
10. After obtaining the values, the parties then again compute the hash codes
11. Party M, compute $\mathbf{Key_M} = J_2^M \bmod J$
12. Party N, compute $\mathbf{Key_N} = J_2^N \bmod J$

Stage 2: Encryption

The proposed model for encrypting and decrypting the text is illustrated in Figure 5.2. The content is translated into ASCII numeric values initially, and subsequently to binary. In addition, the secret key is transformed to binary. These are XNOR'ed after acquiring the binary numbers of both content and secret key. Because the intruder would never understand the key, the XNOR'ed value will make the text complicated and unreadable. Once the XNOR is completed, the obtained unreadable values are shifted one time to the left. These might produce fresh unreadable content. The acquired content is then subjected to 1's complements. The final step entails splitting the interim ciphertext into subsamples that switch locations between each other. The output cipher is then again converted to ASCII and transferred across destination.

The ciphertext is transformed to binary during the decoding step. The gathered data are utilised to split the content into two sections. The placements of these subgroups are switched around. The decoding is then performed on 1's complement of this data. The content obtained in the previous stage is then shifted again in the right progressively bit-by-bit. This content will be then XNOR'ed with the secret key created by the Diffie-Hellman approach. As a result of this, the original statement is created again. It isn't feasible to crack this encryption utilizing Diffie-Hellman message

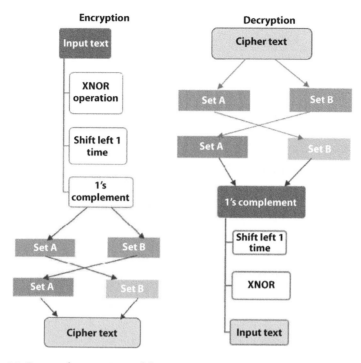

Figure 5.2 Proposed encryption and decryption process.

authentication process. The cypher is made more effective by employing numerous processes. Because the text is XNOR'ed with the secret key, it cannot be broken till the key is shared, as well as the secret key will never be understood until each party's private parameter is known, which again is impossible because it is seldom broadcast across the internet. The procedure is covered in depth underneath. The key is initially produced via Diffie-Hellman and translated to binaries. Encryption is conducted once the key is generated. The approach of encryption and decryption are detailed in Figure. 5.3.

Encryption Procedure:

1. The text to be sent to the destination is decided initially.
2. Each character in the destined text is transmitted to ASCII decimals and then these are transmitted to binary format (8 bits).
3. The resultant output from step 2 is then XNOR'ed with keys of equal length generated by Diffie-Hellman.
4. The resultant output from step 3 is undergone left shift once.

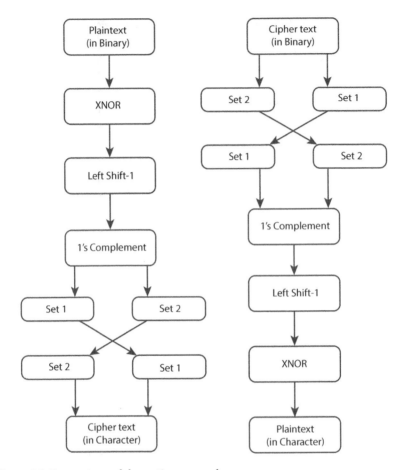

Figure 5.3 Encryption and decryption approach.

5. Then the output from step 4 is undergone one's complement.
6. Then the result is divided into subsets equally (E.g.: 1100 is divided into '11' and '00' subsets). The divided subsets are interchanged from their positions (E.g.: 1100 is interchanged as 0011) and this interchange completely modifies the binary values and the same process enhances the encoded values' complexity.
7. Finally, the result from step 6 is converted back to ASCII decimal and the same is again transmitted to the respective alphabets.

Decryption Procedure

1. The encoded cipher is obtained at the destination.
2. The received cipher is decoded character by character to ASCII decimals initially. Then the result is transmitted to the binary format (8 bits).
3. Then the resultant output is splitted into two halves equally. (E.g.: 1011 is splitted to 10 and 11). These splitted subsets are interchanged with respect to their positions. (E.g.: 1011 is changed to 1110).
4. Final result is undergone one's complement.
5. Then the one's complement output is shifted right once.
6. The result from step 5 is XNOR'ed with the receiver's secret key.
7. Finally, the values are converted back to ASCII decimal and then to binary.

Though the working procedure of ASEEA algorithm is simple, the complexity lies in the key generation phase utilizing hash codes. As the hashing appends some data to the original text, the length of the original text becomes unpredictable to the hackers. ASEEA algorithm will diminish the computation time as well as response time and so the proposed algorithm is simple compared to the others.

Stage 3: LEACH technique

LEACH is proposed for WSNs to offer safe data transit amongst communication parties. LEACH provides encrypted connection amongst sensors while also protecting them from different assaults. The LEACH technique is used for routing in this study. The technique employs a clustering algorithm in which cluster centres are chosen at random. The transmission begins when the cluster leader is chosen. These stages demonstrate how interaction among two nodes is formed by employing LEACH standard and thus are explored in depth below.

1. The sender node transmits the data/secret key to its neighbour cluster heads.
2. Then the cluster head transmits the same to the sink.
3. The sink then again transfers the same to the cluster head of the receiver.
4. The cluster head at the receiver end transmits to receiver.

These steps are repeated. The same process is employed for transmitting the key as well. The various advantages of choosing LEACH are that it completely reduces the data traffic in the entire transmission of both shared key as well as the encrypted data. As routing takes place at a single hop amongst the cluster head and the other nodes, energy is completely saved. Along with that network lifetime is greatly enhanced. Node's location information is not gathered while forming clusters; this in turn increases the privacy. Finally, LEACH operates completely autonomously without any control from the base station so it is distributed in nature.

5.5 Results and Discussion

For evaluating the results, multiple parameters such as time for encryption, time for decryption, response time and cost for computation are considered. For simulation, the LEACH standard is utilized for routing. Five to 1000 nodes for more than 10 rounds are used. MATLAB is employed for simulation.

Environmental setup

- Platform: MATLAB 2021
- Proposed Model: LEACH for routing among WSN
- Simulation area: 100 X 100
- Number of nodes: maximum – 1000, minimum – 5
- Key size: maximum – 2^{10}, minimum – 2^5
- Text size: maximum – 2^{10}, minimum – 2^5

Computational time
The attribute determines how long an algorithm assumes to process a given quantity of data. The outcomes of the suggested approach are contrasted to the research findings of Singh *et al.* (2017). Their research revealed the effectiveness of their suggested technique where key creation, encrypting, and decoding are conducted on input of 10-byte size, as well as the key changes in relation to the various ECC values created. Our suggested solution has been evaluated with three various data sizes, as well as the key size remaining the same while maintaining security. The suggested technique outperforms the research strategy in Singh *et al.*'s investigation by updating the key at each round. As the secret key is produced using the MDH technique, it is extremely challenging to crack because neither party's secret

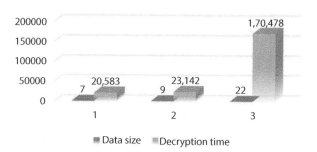

Figure 5.4 Computational time for the input data.

key is ever sent across the internet and arbitrary numbers are utilised to establish generic attributes for public key. Figure 5.4 details the computation time of the ASEEA algorithm and the same is generated by utilizing the data in Table 5.1.

Time for Generating the Key

This attribute computes how long it takes to produce a key each time. Figure 5.5 details the time for generating the key using ASEEA algorithm and the same is generated by utilizing the data in Table 5.1. The calculated time is computed in terms of nanoseconds. Diffie-Hellman is employed to generate unique keys. Various key sizes are assessed with respect to how long it takes to process different sizes of data. The data size of 10 holds a key generation time of around 37,334 nanoseconds, that is appropriate for

Table 5.1 Result analysis of the proposed technique.

Raw input	Parameters considered			Time				
	Message size (bytes)	Length of the key (bytes)	Private key	Generation of key (ns)	Time for encryption (ns)	Time for decryption (ns)	Computational time
Welcome	7	5	Same as above	35,383	15,540	20,583	77,966
Hello Guys	9 (with space)	5	Same as above	37,448	19,241	23,142	83,916
Pleasant Day	22 (with space)	5	Same as above	56, 988	212, 650	170, 478	438, 511

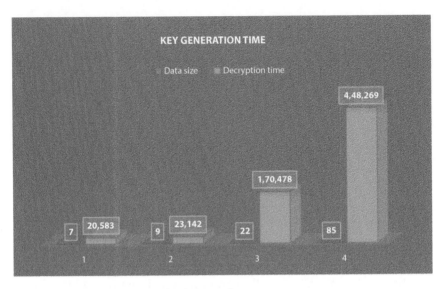

Figure 5.5 Key generation time for the input data.

the WSN context. The comparison of the enhanced Diffie-Hellman algorithm with the existing Diffie-Hellman algorithm is represented in Table 5.2. From the results, it is clearly understood that the enhanced version of the existing algorithm outperforms in terms of computation time and key generation time.

Table 5.2 Comparison of existing and proposed Diffie-Hellman algorithm.

Raw input	Parameters considered			Time				
	Message size (bytes)	Length of the key (bytes)	Private key	Generation of key (ns)		Computational time		
				Existing Diffie-Hellman	Proposed ASSEA	Existing Diffie-Hellman	Proposed ASSEA	
Welcome	7	5	Same as above	30,501	35,383	73,856	77,966	
Hello Guys	9 (with space)	5	Same as above	33,521	37,448	79,961	83,916	
Pleasant Day	22 (with space)	5	Same as above	51,982	56,988	424,514	438,511	

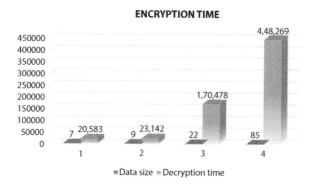

Figure 5.6 Encryption time for the input data.

Encryption Time

This attribute computes the time necessary to execute the suggested cryptographic operations. The technique that consumes the least amount of time to encrypt is termed effective. This suggested encryption technique is evaluated for various data sizes with respect to time required for each input. Figure 5.6 compares the duration of time required to encrypt with respect to data length. If indeed the input is 15 bytes long, the suggested encryption algorithm will require approximately 59,000 ns to encode. 59,000 in seconds is 5.9E-5, which is extraordinarily quick. So, in general the proposed method provides a time complexity of $O(e^n)$ where n is the number of bytes of data. Because the suggested solution requires a relatively short time to encode, it could be regarded as a superior technique for the WSN context.

Decryption Time

This attribute computes the amount of time it takes for a text to create original text from the acquired ciphertext. This suggested approach is evaluated for varied amounts of content in terms of how much time it requires to decode each input if the key is already known. Figure 5.7 compares the duration of time required to decrypt with respect to data length. If indeed the input is 45 bytes long, the suggested technique will require roughly 122,990 ns to decode. The time period is only feasible if the key is shared; without having the key, decryption is impossible. As the number of bytes rises, so does the time required for both encryption and decryption. The suggested techniques require less time than the method presented by Singh et al., 2017. As the size of the data increases, the time for decryption also

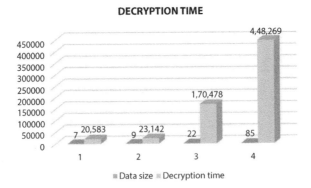

Figure 5.7 Decryption time for the input data.

increases gradually. So, in general the proposed method provides a time complexity of $O(e^n)$. As seen in Figure 5.7, the proposed model greatly enhances the security with minimal encryption time and decoding.

5.6 Analysis of Various Security and Assaults

Numerous measures are employed to assess the trustworthiness of an algorithm across multiple assaults. When dealing with cryptanalysis, numerous assaults on the internet might occur, which must be considered when delivering a safe cryptography strategy.

Plaintext Assault
When the adversary has accessibility to either the plaintext or the ciphertext, this assault happens. This is viewed as a relatively elementary assault on a cryptographic system. The hacker can grab parts of plaintext whenever the transmitter submits content for encryption. The password is never revealed to the hacker since it is sent through a protected network. The hacker attempts to build the encryption scheme, which is then utilised for ciphertext decoding, using some existing cypher and plaintext. The suggested technique does not communicate plaintext across the internet, but only the cipher content is transmitted. Because the key is not really communicated across the internet, this assault becomes extremely challenging to carry out. Even if the assailant acquires any ciphertext, this assault is still hard to execute because the key and cyphertext is changed at every round.

Ciphertext Assault
A known-ciphertext (COA) assault is a cryptographic assault paradigm in which the assailant is assumed to only possess access to a subset of ciphertexts. The assault is accomplished if the associated plaintexts or the key can be guessed. This assault is extremely challenging to carry out because, while ciphertext is delivered over the internet, the secret key is never transmitted across the system. Instead of employing a secure communication, the suggested solution exchanges the key via common metrics. Because a protected channel is not utilised to convey the secret key, the key is unknown to the assailant, making decryption of the ciphertext incredibly hard. The ciphertext cannot be decoded without the secret key, hence there is a good possibility that this assault will not happen if the suggested approach is followed. If the secret key can be retrieved, the ciphertext may be analysed. It is extremely challenging to retrieve the secret key while utilising the recommended technique, making it extremely impossible for this assault to succeed.

Related-key assault
A related-key assault is a type of cryptology in which the assailant may watch the functioning of a cypher under multiple distinct keys whose contents are originally undisclosed, while the assailant is aware of some statistical connection linking the keys. For instance, the assailant may be aware that the final 80 bits of the secret keys are usually identical, even if they never know what bits of data occur initially. Because the suggested solution does not employ the same key for every round of the LEACH, this assault gets more challenging to carry out. If the assailant knows the key, it will only function for a single encrypted text because a new key is created at each round.

Man-in-the-middle assault
Three people are involved in a MIM assault. There would be the client, the person with whom the victim is attempting to connect, and the "guy in the midst," who is interfering with the victim's interactions. Diffie-Hellman is incredibly susceptible to MIM. This exploit will reduce protection by gaining access to all secret attribute values. In customized Diffie-Hellman, hashing is utilised to protect the suggested technique from this assault. This suggested method is safe from MIM attacks after altering Diffie-Hellman with a hash code. Because only two interacting parties can create the right hash value, this assault becomes irrelevant in the suggested technique.

5.7 Conclusion

A WSN is a distributed network that incorporates many circulating, self-focused, compact, low-motorized units known as sensors. Although sensor networks are widely utilised, they are quite complex owing to the restricted amount of power and storage they can consume. Data protection is a major key challenge with WSNs, along with many others, because data going through the internet is never secure as several intruders can gain access to it. The information must be safeguarded from the adversary; therefore, it has been encrypted and converted into an unreadable format. Multiple ways have been employed to protect the data for privacy; however, owing to weaknesses, these techniques are not regarded viable for WSNs. A secured and simple data encryption method is suggested. This strategy will use less computing time. Because the suggested technique has a shorter processing and reaction time, it is the greatest match for WSN cybersecurity. The suggested method also resists plaintext, known ciphertext-only, related-key, and MIM assaults.

References

Akyildiz IF, Su W, Sankarasubramaniam Y, et al. Wireless sensor networks: a survey. *Comput Netw* 2002; 38(4): 393–422.

Elhoseny M, Elminir H, Riad A, et al. A secure data routing schema for WSN using elliptic curve cryptography and homomorphic encryption. *J King Saud Univ-Comput Inf Sci* 2016; 28(3): 262–275.

Mahmood K, Arshad J and Chaudhry SA. An enhanced anonymous identity-based key agreement protocol for smart grid advanced metering infrastructure. *Int J Commun Syst* 2019; 32(16): e4137.

Ullah I, ul Amin N, Iqbal J, et al. An efficient secure protocol for wireless sensor networks based on hybrid approach. *IJCSNS* 2018; 18(6): 59.

Tiwari HD and Kim JH. Novel method for DNA-based elliptic curve cryptography for IoT devices. *ETRI J* 2018; 40(3): 396–409.

Wang W, Si M, Pang Y, et al. An encryption algorithm based on combined chaos in body area networks. *Comput Electr Eng* 2018; 65: 282–291.

Ahmad M, Al Solami E, Wang X-Y, et al. Cryptanalysis of an image encryption algorithm based on combined chaos for a BAN system, and improved scheme using SHA-512 and hyperchaos. *Symmetry* 2018; 10(7): 266.

Chaudhry SA, Shon T, Al-Turjman F, et al. Correcting design flaws: an improved and cloud assisted key agreement scheme in cyber physical systems. *Comput Commun* 2020; 153: 527–537.

Mansoor K, Ghani A, Chaudhry SA, et al. Securing IoT based RFID systems: a robust authentication protocol using symmetric cryptography. *Sensors* 2019; 19(21): 4752.

Somsuk K and Sanemueang C. The new modified methodology to solve ECDLP based on brute force attack. In: *International conference on computing and information technology, Chiangmai, Thailand, 5-6 July 2018*, pp. 255–264. Cham: Springer.

Santhosh R and Shalini M. Security enhancement using chaotic map and secure encryption transmission for wireless sensor networks. *Int J Eng Technol* 2017; 9(2): 689–694.

Daemen J. Limitations of the Even-Mansour construction. In: *International conference on the theory and application of cryptology and information security, Gold Goast, QLD, Australia, 13-16 December 1991*, pp. 495–498. Berlin: Springer.

Singh SR, Khan AK and Singh TS. A new key management scheme for wireless sensor networks using an ellipstic curve. *Indian J Sci Technol* 2017; 10(13): 1–7.

Wiener MJ and Zuccherato RJ. Faster attacks on elliptic curve cryptosystems. In: *International workshop on selected areas in cryptography, Kingston, ON, Canada, 17-18 August 1998*, pp. 190–200. Berlin: Springer.

Neves S and Tibouchi M. Degenerate curve attacks: extending invalid curve attacks to Edwards curves and other models. *IET Inf Secur* 2017; 12(3): 217–225.

Viswanathan S and Kannan A. Elliptic key cryptography with Beta Gamma functions for secure routing in wireless sensor networks. *Wirel Netw* 2019; 25: 4903–4914.

Ghani A, Mansoor K, Mehmood S, et al. Security and key management in IoT based wireless sensor networks: an authentication protocol using symmetric key. *Int J Commun Syst* 2019; 32(16): e4139.

Reeds J. 'Cracking' a random number generator. *Cryptologia* 1977; 1(1): 20–26.

Bogdanov A, Khovratovich D and Rechberger C. Biclique cryptanalysis of the full AES. In: *International conference on the theory and application of cryptology and information security, Seoul, Korea, 4-8 December 2011*, pp. 344–371. Berlin: Springer.

Suresh H and Hegadi RS. DCA-SNC: dual cryptosystem architecture for secure network communication. *Int J Adv Res Comput Sci Software Eng* 2017; 7(1): 198–205.

Nitaj A, Ariffin MRK, Nassr DI, et al. New attacks on the RSA cryptosystem. In: *International conference on cryptology in Africa, Marrakesh, Morocco, 28-30 May 2014*, pp. 178–198. Cham: Springer.

De Weger B. Cryptanalysis of RSA with small prime difference. *Appl Algebra Eng Commun Comput* 2002; 13(1): 17–28.

Roy D and Das P. A modified RSA cryptography algorithm for security enhancement in vehicular ad hoc networks. In: Mandal J, Saha G, Kandar D, et al.

(eds.) *Proceedings of the international conference on computing and communication systems*. Singapore: Springer, 2018, pp. 641–653.

Manger J. A chosen ciphertext attack on RSA optimal asymmetric encryption padding (OAEP) as standardized in PKCS# 1 v2. 0. In: *Annual international cryptology conference, Santa Barbara, CA, 19–23 August 2001*, pp. 230–238. Berlin: Springer.

Babu SS and Balasubadra K. Revamping data access privacy preservation method against inside attacks in wireless sensor networks. *Clust Comput* 2019; 22: 65–75.

Lindell Y and Katz J. *Introduction to modern cryptography*. London: Chapman and Hall/CRC, 2014.

Haakegaard R and Lang J. The Elliptic Curve DiffieHellman (ECDH), http://koclab.cs.ucsb.edu/teaching/ecc/project/2015Projects/Haakegaard+Lang.pdf.

Li J. A symmetric cryptography algorithm in wireless sensor network security. *Int J Online Eng* 2017; 13(11): 102–110.

Khan A, Shah SW, Ali A, et al. Secret key encryption model for wireless sensor networks. In: *2017 14th international Bhurban conference on applied sciences and technology (IBCAST), Islamabad, Pakistan, 10–14 January 2017*, pp. 809–815. New York: IEEE.

Vangala A and Parwekar P. Enhanced encryption model for sensor data in wireless sensor network. In: *2017 20th international symposium on wireless personal multimedia communications (WPMC), Bali, Indonesia, 17–20 December 2017*, pp. 16–21. New York: IEEE.

Albassal EMB and Wahdan A-M. Genetic algorithm cryptanalysis of the basic substitution permutation network. In: *2003 IEEE 46th midwest symposium on circuits and systems, Cairo, 27–30 December 2003*, pp. 471–475. New York: IEEE.

Vangala A and Parwekar P. Encryption model for sensor data in wireless sensor networks. In: Bhateja V, Nguyen B, Nguyen N, et al. (eds.) *Information systems design and intelligent applications*. Singapore: Springer, 2018, pp. 963–970.

Al-Mashhadi HM, Abdul-Wahab HB and Hassan RF. Data security protocol for wireless sensor network using chaotic map. *Int J Comput Sci Inf Secur* 2015; 13(8): 80.

Nidarsh MP and Devi MGP. Chaos based secured communication in energy efficient wireless sensor networks. *Chaos* 2018; 5(6): 742–747.

Sobhy MI and Shehata A-E. Methods of attacking chaotic encryption and countermeasures. In: *2001 IEEE international conference on acoustics, speech, and signal processing, Salt Lake City, UT, 7–11 May 2001*, pp. 1001–1004. New York: IEEE.

Li L, Liu B and Wang H. QTL: a new ultra-lightweight block cipher. *Microprocess Microsyst* 2016; 45: 45–55.

Patil J, Bansod G and Kant KS. LiCi: a new ultralightweight block cipher. In: *2017 International conference on emerging trends innovation in ICT (ICEI), Pune, India, 3-5 February 2017*, pp. 40-45. New York: IEEE.

Maity S, Sinha K and Sinha BP. An efficient lightweight stream cipher algorithm for wireless networks. In: *Wireless communications and networking conference (WCNC), San Francisco, CA, 19-22 March 2017*, pp. 1-6. New York: IEEE.

Solomon M and Elias EP. Data security and privacy in wireless sensor devices. Networks 2018; 5(5).

Kocher PC. Cryptanalysis of Diffie-Hellman, RSA, DSS, and other systems using timing attacks (Extended Abstract) (1995). In: *Annual international cryptology conference, Santa Barbara, CA, 18-22 August 1996*. Berlin: Springer.

Praveena A. Achieving data security in wireless sensor networks using ultra encryption standard version #x2014; IV algorithm. In: *2017 international conference on innovations in green energy and healthcare technologies (IGEHT), Coimbatore, India, 16 March 2017*, pp.1-5. IEEE.

Li C, Li S, Alvarez G, et al. Cryptanalysis of two chaotic encryption schemes based on circular bit shift and XOR operations. *Phys Lett A* 2007; 369(1-2): 23-30.

Ananda Krishna B, Madhuri N, Rao MK, et al. Implementation of a novel cryptographic algorithm in wireless sensor networks. In: *2018 conference on signal processing and communication engineering systems (SPACES), Vijayawada, India, 4-5 January 2018*, pp. 149-153. New York: IEEE.

Praveena A and Smys S. Efficient cryptographic approach for data security in wireless sensor networks using MES VU. In: *2016 10th international conference on intelligent systems and control (ISCO), Coimbatore, India, 7-8 January 2016*, pp. 1-6. New York: IEEE

Kumar, P. S., Karthiga, M., & Balamurugan, E. Cyber ML-Based Cyberattack Prediction Framework in Healthcare Cyber-Physical Systems. In Computational Intelligence in Robotics and Automation (pp. 141-159). CRC Press.

Gracy PL and Venkatesan D. An honey encryption based efficient security mechanism for wireless sensor networks. *Int J Pure Appl Math* 2018; 118(20): 3157-3164.

Yue T, Wang C and Zhu Z. Hybrid encryption algorithm based on wireless sensor networks. In: *2019 IEEE international conference on mechatronics and automation (ICMA), Tianjin, China, 4-7 August 2019*. New York: IEEE.

Sountharrajan, S., Suganya, E., Karthiga, M., Nandhini, S. S., Vishnupriya, B., & Sathiskumar, B. (2020). On-the-Go Network Establishment of IoT Devices to Meet the Need of Processing Big Data Using Machine Learning Algorithms. In *Business Intelligence for Enterprise Internet of Things* (pp. 151-168). Springer, Cham.

Karthiga, M., Nandhini, S. S., Tharsanee, R. M., Nivaashini, M., & Soundariya, R. S. (2021). Blockchain for Automotive Security and Privacy with Related Use

Cases. In *Transforming Cybersecurity Solutions Using Blockchain* (pp. 185–214). Springer, Singapore.

Kumar, G. S., Premalatha, K., Aravindhraj, N., Nivaashini, M., & Karthiga, M. Secured Cryptosystem Using Blowfish and RSA Algorithm for the Data in Public Cloud. *International Journal of Recent Technology and Engineering (IJRTE)* ISSN, 2277–3878.

Sountharrajan, S., Nivashini, M., Shandilya, S. K., Suganya, E., Bazila Banu, A., & Karthiga, M. (2020). Dynamic recognition of phishing URLs using deep learning techniques. In *Advances in cyber security analytics and decision systems* (pp. 27–56). Springer, Cham.

Shree, S. I., Karthiga, M., & Mariyammal, C. (2017, January). Improving congestion control in WSN by multipath routing with priority based scheduling. In *2017 International Conference on Inventive Systems and Control (ICISC)* (pp. 1-6). IEEE.

Lasry G, Kopal N and Wacker A. Cryptanalysis of columnar transposition cipher with long keys. *Cryptologia* 2016; 40(4): 374–398.

Prusty AR. The network and security analysis for wireless sensor network: a survey. *Int J Comput Sci Inf Technol* 2012; 3(3): 4028–4037.

Diffie W and Hellman M. New directions in cryptography. *IEEE T Inform Theory* 1976; 22(6): 644–654.

6

Security and Confidentiality Concerns in Blockchain Technology: A Review

G. Prabu Kanna[1]*, Abinash M.J.[2], Yogesh Kumar[3], Jagadeesh Kumar[4] and E. Suganya[5]

[1]School of Computing Science and Engineering, VIT Bhopal University, Madhya Pradesh, India
[2]Department of Computer Science and Engineering, SOT, Pandit Deendayal Energy University, Gujarat, India
[3]Department of Information Technology/PG (CS), AKCAS, Srivilliputhur, India
[4]Department of Computer Science and Engineering, Kathir College of Engineering, Tamil Nadu, India
[5]Department of Information Technology, Sri Sivasubramaniya Nadar College of Engineering, Kalavakkam, Chennai, Tamil Nadu, India

Abstract

Today's culture is talking a lot about blockchain, which is rapidly gaining popularity. Despite the fact that it has already altered many people's lifestyles in certain ways, opponents have highlighted worries about its scalability, security, and sustainability due to its tremendous impact on several sectors and enterprises. Numerous industries, including banking, healthcare, transportation, risk management, the Internet of Everything (IoE), as well as social and public services, have adopted blockchain, which is a decentralised technology. It takes considerable ability to resolve commercial issues. Each transaction on a blockchain is connected to prior transactions or records, and the records are encrypted. Algorithms running on nodes verify transactions happening through blockchain. One individual (or) entity could not start a transaction. Ultimately, blockchains offer transparency, enabling any user to keep track of transactions at any moment. In this chapter, we try to perform an in-depth analysis of blockchain technology by looking at the scenarios and problems from the privacy and security viewpoints.

---

*Corresponding author: gpkanna@gmail.com

S. Sountharrajan, R. Maheswar, Geetanjali Rathee, and M. Akila (eds.) Wireless Communication for Cybersecurity, (129–148) © 2023 Scrivener Publishing LLC

Keywords: Blockchains, security issues, confidentiality concerns, advancements and future scope

6.1 Introduction

Blockchain technology has enormous potential for a wide range of applications and offers numerous opportunities for various infrastructure. Resource management is encouraged by the technology, which also guarantees effective and safe communication [1]. One question that is becoming increasingly common is about the cryptocurrency bitcoin and the technology source that powers it, known as blockchain. Confidence is raised when parties carry out banking transactions with blockchain since it lowers the risk of theft and instantly creates a record of activity, thus establishing an automated background investigation for each system user. Because of its decentralised nature, blockchain creates dependability and lowers the risk involved in entering into a business agreement with an unfamiliar party.

Blockchain technology is crucial to the advancement of industry. The development of privacy protocols and blockchain decentralisation technologies for safeguarding, data services, auditing, and regulating transactions on digital platforms might be advantageous to many businesses. Blockchain is based on decentralised and secure distributed protocols without a central authority or source of control, and data blocks are created, added to, and confirmed by network nodes themselves. Blockchain is a distributed ledger that utilises cryptographic algorithms [2, 3]. Images, texts, video calls, and voice conversations may all be made and received straight over the internet. The sender and the recipient must retain a trusted third party throughout the transaction. In the conventional system, consumers have to depend on a third party to execute their financial activity. On the other hand, blockchain will offer complete transactional protection. Every transaction should be recorded in a block, which will behave as a record book. Every time a transaction is finished, a block is appended to the blockchain, which serves as a permanent database. When a current block is finished, a new block is usually added or generated. Every block contains a hash of the block before it. A distributed peer-peer network is utilized to manage bitcoin, the first decentralised digital currency ever created. Bitcoin was subsequently acknowledged as the leading currency with respect to user acceptance and broad use [4]. The blockchain's primary function is to record time-stamped data from transactions in data blocks that are linked together in a chain in the order in which they occurred. Every block is given a distinct hash value through a cryptographic procedure to ensure the confidentiality of the data. Similar to a linked list, these hash values serve as linkages between these blocks. Asset owners can use blockchain to track and trade valuable items, such as outstanding invoices, in a secure, transparent, private, and self-reconciling "chain" of transactions. Every block contains the hash of the previous block, which makes it easier to link the blocks that make up the blockchain [5]. Blockchain-based networks took a while to catch on because of their complex design, but eventually a variety of global businesses, including the financial, healthcare, logistics, manufacturing, and energy sectors [6] started to pay attention to them. The blockchain operates inside a sophisticated framework that brings together a variety of other widely used technologies, such as distributed environments, decentralised architecture, peer-to-peer networking, encryption and others [7]. Blockchain technology has been employed in areas other than digital currency, such as the Internet of Things. Blockchain technology could be used in many everyday business-to-business transactions in the future, including those powered by enterprise applications [8,9]. Autonomous marketplaces for other assets are likely to proliferate. Because the software is a controlled and open framework that is visible to all transaction participants, a blockchain-based transaction eliminates the need for third-party oversight. Blockchain technology is expected to have huge implications when it becomes more widely used, radically changing how people use the internet [10-12]. On a commercial level, it is being adopted by enabling advances in the IT businesses for the enhanced efficacy and streamlined business activities. In order to support the growth of businesses and carry out independent blockchain research, key industry players like Google and Microsoft have built outlines that gives blockchain architecture-based service to clients.

6.2 Blockchain Technology

Bitcoin allows for decentralised peer-to-peer exchange of digital cash via the internet. As a distributed ledger that is open to all users and is hosted by several willing hosts, or nodes, the blockchain is implemented in bitcoin. Blockchain is a method of recording information that makes it impossible or difficult to change, hack, or manipulate the system. A blockchain, like a database, stores information electronically in digital format. Blockchains are best known for their critical role in cryptocurrency systems like bitcoin, where they keep a secure. A blockchain is a distributed ledger that duplicates and distributes transactions across the blockchain's network of computers. The ledger authenticates and records transactional data sent via the network. Given that the whole network, as opposed to a single authority as in traditional financial frameworks, is responsible for the verification of the transactions, the capacity of the blockchain in bitcoin to prevent double spending in transactions is a distinguishing feature. In its most basic form, a blockchain is a ledger that securely extends the chain over time while recording transactions in an immutable, append-only manner.

Cryptographic methods are used to protect the blocks within the blockchain, ensuring the integrity of the transactional information [13]. The integrity of the data is ensured by the permanent records that make up the blockchain, which cannot be changed or tampered with. The dispersed nodes of the network are used to transit the data on the blockchain. The blockchain is distinct from previous technologies in that it provides an overall order of blocks by timestamping data entries. The data hash value inside a block and the hash value of the block preceding it serve to link the chain of blocks together. The existing blockchain can only be expanded with a new block if the consensus technique has been properly applied. This consensus procedure must control chain entry rights, follow security protocols for block verification, and guarantee record consistency across all network nodes. As is obvious, the blockchain is a distributed ledger that uses a decentralised network to safely and impartially verify all financial information. Security concerns including data breaches or transactions, reliance on a third party or centralized body, and the unpredictability of other parties may be addressed by a blockchain-based system. Blockchain technology has developed significantly and undergone an evolution; Figure 6.1 below illustrates these phases.

The first iteration of the blockchain technology, known as version 1.0, had a decentralized public ledger for holding money. Version 2.0 of the blockchain includes a system for maintaining trust through payment systems that runs independently without the involvement of any other parties. The third stage, known as blockchain version 3.0, represents the technology's present and future. It encompasses a number of application domains,

Figure 6.1 Blockchain development stages.

including distributed finance issues, IoE, identity managing, education, big data, artificial intelligence, healthcare, as well as others [14].

6.3 Blockchain Revolution Drivers

Blockchain is a fundamental innovation that paves the path for the change in thinking from believing people to relying on machines although from central to distributed authority, in addition to its cutting-edge architecture and applications. To fully grasp blockchain's potential, we may study it from two angles. The ability to keep track of who owns what assets on and off platforms, as well as the rights and duties that come with contracts, may initially form an Information and Communications Technology (ICT).

A blockchain may be used to store any kind of data, including ownership of assets, contractual responsibilities, copyrights for creative works, credit exposures, and digital identities. Second, blockchain may be considered a technical institution that decentralises the governance frameworks that serve as the foundation for social and economic decision-making. While we use an ICT viewpoint, the major forces behind the blockchain revolution may be explained from both an institutional and an ICT standpoint. These elements are illustrated in Figure 6.2 and are explained as follows.

6.3.1 Transparent, Decentralised Consensus

The sequence in which applications, activities (deploy and invoke), and data have been performed, updated, or produced is confirmed via a blockchain technique known as consensus. The appropriate sequence is crucial because it can produce ownership, which can lead to rights and duties. Blockchain networks are decentralised, meaning there is no central hub or authority that decides what happens when, accepts transaction activities, or creates rules that allow nodes to communicate with each other.

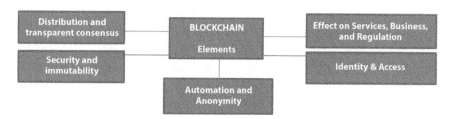

Figure 6.2 Blockchain elements.

6.3.2 Model of Agreement(s)

The consensus model or models aid in maintaining the veracity of data stored on blockchains. Issues about different consensus mechanisms, which include blockchain splits, consistency breakdowns, dominance issues, verifying nodes, and inferior network functionality, may surface when a consensus method fails. A consensus method has the following three qualities based on practicality and efficacy:

1. Safety - For a consensus protocol to be secure and dependable, each node must produce output that complies with the protocol's specifications.
2. Liveness - In order to offer a value, a consensus method makes sure that all active, non-faulty nodes are present.
3. Fault Tolerance - A compromising protocol allows tolerating failures simultaneously permitting a fault node to participate in the protocol recovery.

6.3.3 Immutability and Security

A shared, vandal duplicated record referred to as a blockchain uses one-way cryptographic algorithms to make records unchangeable and nonrepudiable. Records are made irreversible and nonrepudiable using one-way cryptographic hash techniques. A reliable historical database that has received universal support assists to increase confidence in the system. If someone or any entity doesn't have influence over most of the miners, it gets very hard for anyone to mess with the record (voters). Blockchain has been alluded to as the "trust machine" by *The Economist*.

6.3.4 Anonymity and Automation

Through blockchain, a collection of individuals may work together while accessing global data sources, with automated reconciliation between all contributors. Using public/private key technology, the owning claims to the material are performed and data transfers are permitted, eliminating the need for personal communication, trust providers, validation, or adjudication. The application makes sure that duplicated or conflicting data cannot be entered to the ledger indefinitely. Automation is the use of algorithms (smart contracts or smart contract software) to monitor, evaluate, and control the implementation of contracts in an automated manner.

Blockchain technology's basic goal is to seamlessly transfer confidence from a particular central body to the entire network. The blockchain network's nodes each have the capacity to safely send and update data. A decentralised autonomous organisation (DAO), also known as a decentralised autonomous corporation, is an organisation that abides by some protocols delivered as computer programmes and known as smart contracts (DAC). Blockchain uses blocks to record the details of smart contracts and transaction history.

6.3.5 Impact on Business, Regulation, and Services

Both the public and commercial sectors have high hopes for blockchain technologies since they are the cornerstone for building peer-to-peer networks for exchanging data, assets, and digital goods without middlemen. Blockchain could immensely improve the application of governance and regulatory controls across a variety of economic sectors and in whole unique ways. In the framework of the present fourth industrial revolution that is characterised by fusing numerous technologies that distort the barriers between physical and virtual space, blockchain is a component of a wider toolset. Blockchain has the potential to upend numerous businesses and society as a whole when paired with other cutting-edge technologies like AI, driverless cars, fog computing, and machine learning.

6.3.6 Access and Identity

Three crucial factors—public (or) without permission, allowed permission (or) private, and consortium—affect a blockchain's identification and function. In-depth discussion of these blockchain criteria may be found in [15]. On a private blockchain, users have fewer options for creating smart contracts and validating block transactions. This is suited for normal enterprises and governance structures [16]. Public blockchains are designed to keep the degree of security while eliminating the middleman from transactions [19]. Anyone with internet connection and a public blockchain may join the network by creating smart contracts and participating in block verifications.

6.4 Blockchain Classification

Despite the fact that the structure, accessibility, and verification of blockchain technology are continually evolving, many application areas are

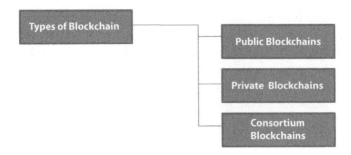

Figure 6.3 Types of blockchain.

embracing it. Users can select from the three types of blockchains described below based on their demands and the circumstance [20-22]. All of these different blockchain models have certain essential characteristics in common, like a decentralized architecture, interacting among peer-peers, consensus operations as well as timestamping, while they differ from one another. Public, private, and consortium blockchains are the three different types of blockchains that are shown in Figure 6.3.

6.4.1 Public Blockchain

Accessing data using a blockchain's distributed, open network is not subject to any restrictions. Nevertheless, a public blockchain may be either written with permissions or without them. In a permissioned network, only a select few nodes are authorized to perform new transactions (which are recorded in the blockchain), verify the purchases made by other nodes, and review the transaction log. If the network has no permission limitations, anybody can write into it (reading the blockchain). Real evidence consensus enhances the trustworthiness of the public blockchain. Since a large number of nodes often join the network as soon as it is made accessible to the public, and more nodes equal a more dispersed network, such a blockchain is viewed as safe. Additionally, the blockchain is transparent since all nodes may view the records ledger. However, there are several limitations like the poor processing speed caused by the network's many nodes. Since proof-of-work needs a lot of time and labour to verify requests, such blockchains have problems with scalability and efficiency. The most popular public blockchains on the market right now are Bitcoin [3], Litecoin [17], and Ethereum [18].

6.4.2 Private Blockchain

A private blockchain's structure includes a few restrictions and operates in a closed system. These solutions are preferred when a company wants a blockchain with access and involvement from a small number of users. Additionally, no one is allowed to see the information or be involved in transaction activities inside the blockchain [21]. Such a thing may be used to secure consumer resources, monitor supply chains for artificial intelligence-based people, etc. It is managed by the enterprises themselves. These blockchain networks might be either permission is allowed or not inside the private group of users. Private blockchains perform computations more rapidly than public blockchains because to the lower number of participating nodes, and they are scalable, allowing for the option to change the number of nodes according to demand. As a result, transaction verification and validation are more effectively done. One significant disadvantage of private blockchains is that they don't provide the same degree of decentralised security as public blockchains.

6.4.3 Blockchain Consortium

Partial decentralisation may be demonstrated in the consortium blockchain, which blends public and private blockchains. In this blockchain network, the node has the power to decide in advance whether the data or transaction details are public or private. It's critical to comprehend the differences among a consortium and a fully private blockchain. It is possible to think of this blockchain, also known as federated blockchain, as a publicized blockchain with authority where anybody may access data through the network but only the representative nodes are permitted to put data into the network. Figure 6.4 displays the corresponding pattern

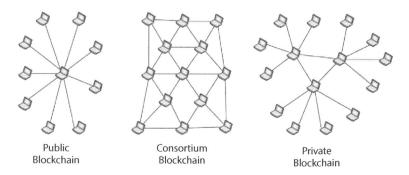

Figure 6.4 Blockchain type pattern representation.

representations. This blockchain features a lot of nodes, much like a public blockchain, but unlike a private blockchain, the nodes are subject to specific restrictions. This kind of blockchain is often used in the financial and government sectors, such as R3, the Energy Web Foundation, etc.

6.5 Blockchain Components and Operation

A distributed network called a blockchain is employed to safely collect financial information logs. As shown in Figure 6.5 below, the blockchain maintains data in the form of chained-together blocks. The blockchain grows in size as more transactions take place because a block is issued after a certain period of time that contains data about the activities that took place during that time. The mining procedure starts when a transaction is requested and broadcasts the request to all network nodes for consensus protocol approval. The block is only appended to the chain when it is verified by every other node.

A blockchain is a growing collection of information known as blocks that are joined together and encrypted. Each block generally contains transaction information, a timestamp, and a cryptographic hash of the preceding block. The adjacent record of transaction blocks may be saved in a tiny repository of data (or) in flat files recognitions to the efficient structure of blockchain data. These blocks are connected together, and each link in the chain refers to the block that came before it. A chain's genesis block is the first one. The blockchain is shown as a vertical stack, with the genesis block at the bottom and blocks placed on top of one another. A lot of information regarding the structure of blockchain is provided in [23, 24]. All blocks are

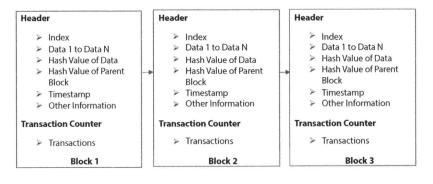

Figure 6.5 Blockchain structure.

said to be uniquely recognised by cryptographic hashes produced by the SHA256 method and stored in the block header.

6.5.1 Data

The services and applications used in a blockchain decide the data kept there. Peer-to-peer file systems like storj, Ethereum swarm, sia, and others might utilise it to store files in the cloud. Applications for the stored data include recording transaction information, banking, contracts, and IoT.

6.5.2 Hash

Let's start by defining a cryptographic hash function as it is essential to both techniques. A cryptographic hash function produces a fixed-size hash from a variable-length input. In other words, it changes a variable-size bit array from an arbitrarily huge input (hash).

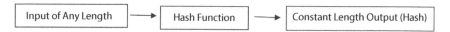

Data integrity checks and file identification frequently involve the usage of cryptographic hash algorithms. Comparing hashes is quicker and simpler than comparing the actual data. Additionally, they are employed for password verification, database storage of private information (such as passwords), and authentication purposes.

6.5.3 MD5

A cryptographic hash function called MD5 creates a 128-bit hash from data that can be any length. Despite being regarded as cryptographically defective, it is nonetheless commonly used in several contexts. Verifying the integrity of files exchanged for PR purposes is one of the most popular uses. The 512-bit data is processed by the MD5 algorithm in 16 words of 32 bits each. There is a 128-bit hash as a result. As we previously noted, the MD5 is regarded as cryptographically broken. Let's discuss its security in more depth. There are potential MD5 assaults. Such methods might result in collisions on a normal computer within a minute. Results have given enough justification to eliminate using MD5 in applications that need collision resistance, including digital signatures. For solutions requiring a high level of security, MD5 is no longer advised. However, it's frequently utilised as a file checksum. One message authentication methodology to verify the content of outsourced files is MD5 [25, 26].

6.5.4 SHA 256

A family of hash algorithms that is extensively used is SHA. SHA-256 ranks among the safest and highly used hashing techniques. To begin with, it is a one-way operation. Therefore, it is quite challenging to deduce the input from the hash. A brute force assault would ideally need 2,256 tries to be successful. Second, SHA-256 is collision-resistant. This is due to the 2,256 different hash values that may be used. As a result, in practise there is essentially little danger of accident. Lastly, the SHA-256 makes use of the avalanche phenomenon. The input may be slightly altered to get a completely new hash. In summary, the cryptographic hash function SHA-256 meets all important requirements. It is widely employed in applications that require a high degree of safety as a consequence.

6.5.5 MD5 vs. SHA-256

First of all, MD5 produces 128-bit hashes. In addition, SHA-256 is safer than MD5, especially when it comes of resistance to collisions. This means that applications using extensive security protocols shouldn't utilise MD5. The SHA-256, on the other hand, is used for high-security operations like SSL handshakes or digital signatures. Furthermore, fewer vulnerabilities against SHA-256 than MD5 have been documented. A normal computer is assumed to be capable to attack the MD5 because to its poor cryptography.

Speed-wise, MD5 is a little bit quicker than SHA-256. As a result, the MD5 checksum is commonly used to confirm the integrity of data. In summary, SHA-256 often performs better than MD5. It is more stable, reliable, and less prone to break. It is not really relevant that SHA-256 is a tiny bit slower than MD5 unless speed becomes the main factor. The longer hash causes the algorithm to run more slowly. As a result, SHA-256 achieves the best balance between security and speed. As a result, SHA-256 typically outperforms MD5, particularly when it comes to safety. On the other hand, systems where speed is the most essential element and where a high degree of safety is not required can use MD5. The SHA-256 algorithm is not the fastest at all. When hashing short strings, SHA-256 is approximately 30% faster than SHA-512. For each case, three measurements were taken, and an average value was determined. The time is in milliseconds per 1 000 000 measures. The equipment makes use of a personal computer (PC) having a 64-bit Windows 10 operating system, with a single Intel i7 2.60GHz core and 16GB of RAM. The UUID means universally unique identifier. A UUID is a value of 128 bits. The results are shown in Table 6.1 below. Figure 6.6 hashing time in ms vs. file size in character

Table 6.1 Time to encrypt (in milliseconds) for file sizes in character.

Data to encode in length	MD5 hash average 1m (ms)	SHA-256 hash average 1m (ms)
50 Character	770	855
75 Character	851	1172
100 Character	1010	1401
150 Character	1611	2186

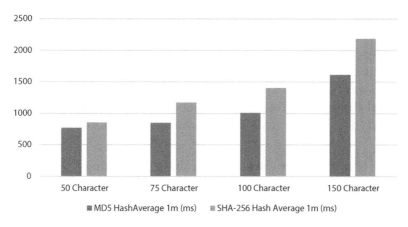

Figure 6.6 Hashing time in ms vs. file size in character.

Timestamp: It is essential to note the moment the block was generated. Tracking the creation or modification time of a document using timestamping is safe. Because it enables the parties to ascertain the source and accessibility of a document at a given moment and date, this approach is quickly becoming a crucial instrument in business.

Additional Information such as nonce, digital signatures, and a few user-defined values are examples of other data. Each user has two keys—a private and a public key. These two keys are needed to create a digital signature that is used for both signing and verifying. The data is encrypted using the private key, which is kept confidential and utilized to approve a transaction through signature. The public key is used to authenticate and decrypt data during the transaction verification phase, hence ensuring data validity. The public key is known to everyone. A 4-byte number that starts with 0 and rises every time a hash computation is made is known as a nonce value. The goal threshold value of a valid block hash is determined by the nbits value, in accordance with [27, 28].

6.6 Blockchain Technology Applications

6.6.1 Blockchain Technology in the Healthcare Industry

Patients are hesitant to discuss medical specifics with strangers in today's society. In this situation, the patient can employ technology to shield all information from prying eyes. A smartphone app or a website can be used to access this blockchain. Health records, papers, and photographs are primarily found in healthcare blocks. [29] discusses the healthcare blockchain in great detail and shows how the data has an impact on both storage and throughput. If data were stored on bitcoin-inspired blockchains, each user would have a copy of every user's health record. Blockchain technology enables users to access a single data source to obtain quick, accurate, and comprehensive healthcare data. It should come as no surprise that safeguarding our sensitive medical information is a top priority.

6.6.2 Stock Market Uses of Blockchain Technology

Blockchain technology reduces the expenses associated with exchanging assets, increases access to global markets, and reduces volatility in the traditional securities market by cutting out the middlemen in the transfer of property rights. It's crucial to keep track of who owns what when people purchase or exchange resources like stocks, mortgages, or commodities. In today's capital markets, there are brokers, exchanges, central security deposits, intermediaries, and banks. Through automation and decentralisation, blockchain can make stock exchanges much more efficient. Blockchain technology has the potential to solve interoperability, trust, and transparency issues in fragmented market systems. Stock market participants like traders, brokers, regulators, and stock exchanges must go through a time-consuming process.

These parties are built on an antiquated, lax, yet ineffective paper ownership system [30]. Blockchain can significantly increase the efficiency of stock exchanges through automation and decentralization. To a large extent, blockchain can eliminate the need for third-party regulators since the rules and regulations are built into smart contracts and enforced with each trade to register transactions, with the blockchain network acting as a regulator for all transactions. Using blockchain and smart contracts in post-trade activities can eliminate the need for intermediaries because peers rather than an intermediary handle transaction confirmations. The expenses associated with intermediaries, such as those for trade record keeping, audits, and trade verifications, increase as the number of intermediaries in the system reduces.

6.6.3 Financial Exchanges in Blockchain Technology

Decentralized bitcoin exchange providers have multiplied in recent years. Blockchain exchanges allow for quicker and more affordable transactions. Investors also have more control and security because a decentralised exchange does not require them to deposit their assets with a centralised authority. Although cryptocurrencies are the main focus of blockchain-based exchanges, the idea could also be applied to more conventional investments. In addition, since a decentralised exchange does not require investors to deposit their funds with a centralised authority, they have more control and security. Although cryptocurrencies are the main focus of blockchain-based exchanges, the idea could also be applied to more conventional investments.

6.6.4 Blockchain in Real Estate

Real estate transactions require a tonne of paperwork to transfer deeds and titles to new owners, verify financial information, and verify ownership. Historically, real estate technology has been focused on listing. Blockchain can help to simplify and secure the process of buying and selling properties. For purchasers, this means being able to research the ownership and history of a property. It entails being able to give sellers greater information about the sale process. The use of blockchain technology to record real estate transactions may offer a more secure and convenient way of verifying and transferring ownership. Blockchain introduces new ways to trade real estate and can enable trading platforms and online marketplaces to more fully support real estate transactions. This may facilitate transactional speed, reduce paperwork, and save money. Blockchain enables the speedier and more efficient completion of real estate transactions. This is due to the fact that blockchain enables digital asset transfers, doing away with the need for paper contracts or other physical documentation.

Additionally, blockchain facilitates secure data sharing, makes it easier to collect and pay rent to property owners, and offers superior due diligence across the portfolio. This increases operational efficiency while saving time and money, and it also generates a lot more data to help with decision-making. Another benefit of blockchain is its high level of security. Since every transaction is recorded on a distributed ledger and is unchangeable, it gives buyers and sellers peace of mind that the process is safe and secure.

6.6.5 Blockchain in Government

The hyper-connectivity that can now be seen in the world around us has resulted not only in more data but also in a significant shift in how the economy is operating and interacting. The government must transform itself so that is truly centred on its citizens by becoming more open, effective, cost-conscious, and real-time in this era of constant change and economic change. And in order to meet this newly discovered demand, a change that would rock a bureaucratic government agency is required. The introduction of a secure blockchain architecture and other aspects of this technology must be encouraged for this shift to occur. There are many advantages of a decentralised government centre; it makes government entities more effective, both in terms of how they operate and in terms of how well-liked they are by the general public. Another application for blockchain-stored digital identities is the administration of government benefits such as welfare programmes, Social Security, and Medicare. Using blockchain technology could reduce fraud and operational costs. Meanwhile, beneficiaries can receive funds more quickly thanks to blockchain-based digital disbursement.

And blockchain offers a solution to this issue with all of its unique features. By enabling users to access and validate data, transparency, the key component of blockchain applications, chnges public attitudes to government. By enabling citizens to independently verify the claims made by the government, blockchain solutions accelerate the entire problem-solving process. When used properly, blockchain technology can reduce costs while also preventing duplication of effort, streamlining workflows, boosting security, lessening the load of audits, and even ensuring that data integrity is preserved. Many government operations, like taxation and voting, may be streamlined with the improved openness and security that blockchain technology can offer. The increased transparency and security that blockchain technology can provide may allow for the simplification of many government processes, including taxation and voting.

6.6.6 Other Opportunities in the Industry

Blockchain can provide access to the banking and payment sectors for billions of users worldwide, including those in third-world countries that do not have access to traditional banking. Blockchain technology is being used by several financial organisations to streamline, improve, and safeguard their processes. Blockchain investments are rising in initiatives and enterprises related to banking and payments.

6.7 Difficulties

Future prospects for blockchain technology include both benefits and difficulties. Although considerable, the difficulties may be solved as technology develops and advances.

Scalability and Security - The blockchain is becoming bigger and bigger as more people use it and there are more transactions taking place every day. In open networks like public blockchains, this is a constant issue. Privacy is reduced in decentralized systems that replicate data throughout their network. Although there are numerous difficulties, the integrity of blockchains is their main strength.

6.8 Conclusion

Blockchain has been a fascinating topic of late, and it will support a wide range of uses. Blockchain will give a greater security during any valued financial transaction. This technology is mainly envisioned to handle bitcoin transactions. Blockchain applications comprise smart contracts, Ethereum, and distributed ledgers. This also increases security. The most appropriate and widely used blockchain application is bitcoin. Their transactions are faster and more economic than any other application. It can improve safety measures, especially for sensitive data. Blockchain applications frequently profit from its transparency and immutability.

References

1. Ashwini, K., Amutha, R., Immaculate, R. R., & Anusha, P. (2019, March). Compressive sensing based medical image compression and encryption using proposed 1-D chaotic map. In *2019 International Conference on*

Wireless Communications Signal Processing and Networking (WiSPNET) (pp. 435-439). IEEE.
2. Aste, T.; Tasca, P.; Di Matteo, T. Blockchain technologies: The foreseeable impact on society and industry. *Computer* 2017, 50, 18–28. [CrossRef]
3. Nakamoto, S. Bitcoin: A Peer-to-Peer Electronic Cash System. 2019. Available online: https://bitcoin.org/bitcoin.pdf (accessed 11 February 2012).
4. Xie, S.; Zheng, Z.; Chen, W.; Wu, J.; Dai, H.-N.; Imran, M. Blockchain for cloud exchange: A survey. *Comput. Electr. Eng.* 2020, 81, 106526. [CrossRef]
5. Lu, Y. Blockchain and the related issues: A review of current research topics. *J. Manag. Anal.* 2018,5, 231–255. [CrossRef]
6. Al-Jaroodi, J.; Mohamed, N. Blockchain in industries: A survey. *IEEE Access* 2019, 7, 36500–36515. [CrossRef]
7. Crosby, M.; Pattanayak, P.; Verma, S.; Kalyanaraman, V. Blockchain technology: Beyond bitcoin. *Appl. Innov.* 2016, 2, 6–10.
8. Zhang, Y.; Wen, J. The IoT electric business model: Using blockchain technology for the internet of things. *Peer-to-Peer Netw. Appl.*2017, 10, 983–994. [CrossRef]
9. Sun, J.; Yan, J.; Zhang, K.Z.K. Blockchain-based sharing services: What blockchain technology can contribute to smart cities. *Financial Innov.* 2016, 2, 26. [CrossRef]
10. Peck, M.E. Blockchains: How they work and why they'll change the world. *IEEE Spectr.* 2017, 54, 26–35. [CrossRef]
11. Idrees, S.M.; Alam, M.A.; Agarwal, P.; Ansari, L. Effective Predictive Analytics and Modeling Based on Historical Data. In *Proceedings of the Advances in Service-Oriented and Cloud Computing, Taormina, Italy, 15–17 September 2015*; Springer: Singapore, 2016; pp. 552–564.
12. Pazhaniraja, N., Sountharrajan, S., Suganya, E. et al. Optimizing high-utility item mining using hybrid dolphin echolocation and Boolean grey wolf optimization. *J Ambient Intell Human Comput* (2022). https://doi.org/10.1007/s12652-022-04488-3
13. Narayanan, A.; Bonneau, J.; Felten, E.; Miller, A.; Goldfeder, S. *Bitcoin and Cryptocurrency Technologies: A Comprehensive Introduction*; Princeton University Press: Princeton, NJ, USA, 2016.
14. Tschorsch, F.; Scheuermann, B. Bitcoin and beyond: A technical survey on decentralized digital currencies. *IEEE Commun. Surv. Tutorials* 2016, 18, 2084–2123.
15. M. Pilkington, Blockchain technology: Principles and applications. In *Research Handbook on Digital Transformations*, edited by F. Xavier Olleros and Majlinda Zhegu. Edward Elgar, 2016.
16. M. Han, J. Li, Z. Cai and Q. Han, Privacy reserved influence maximization in gps-enabled cyber-physical and online social networks, in *2016 IEEE International Conferences on Social Computing and Networking (SocialCom)*, IEEE, 2016, 284–292.

17. Bentov, I.; Lee, C.; Mizrahi, A.; Rosenfeld, M. Proof of activity: Extending bitcoin's proof of work via proof of stake [extended abstract]. *ACM SIGMETRICS Perform. Eval. Rev.* 2014, 42, 34–37. [CrossRef]
18. Wood, G. Ethereum: A secure decentralised generalised transaction ledger. *EthereumProj. Yellow Pap.* 2014, 151, 1–32.
19. Satybaldy, A.; Nowostawski, M. Review of Techniques for Privacy-Preserving Blockchain Systems. In *Proceedings of the 2nd ACM International Symposium on Blockchain and Secure Critical Infrastructure, Taipei, Taiwan, 5 October 2020*; pp. 1–9.
20. I.-C. Lin and T.-C. Liao, A survey of blockchain security issues and challenges., *IJ Network Security*, 19 (2017), 653–659.
21. X. Xu, I. Weber, M. Staples, L. Zhu, J. Bosch, L. Bass, C. Pautasso and P. Rimba, A taxonomy of blockchain-based systems for architecture design, in *Software Architecture (ICSA), 2017 IEEE International Conference on*, IEEE, 2017, 243–252.
22. Z. Zheng, S. Xie, H.-N. Dai and H. Wang, Blockchain challenges and opportunities: A survey, *Int. J. Web and Grid Services*, Vol. 14, No. 4, 2018.
23. Z. Zheng, S. Xie, H. Dai, X. Chen and H. Wang, An overview of blockchain technology: Architecture, consensus, and future trends, in *Big Data (BigData Congress), 2017 IEEE International Congress on*, IEEE, 2017, 557–564.
24. I.-C. Lin and T.-C. Liao, A survey of blockchain security issues and challenges., *IJ Network Security*, 19 (2017), 653–659.
25. Chauhan, M.M. An implemented of hybrid cryptography using elliptic curve cryptosystem (ECC) and MD5. In: *IEEE International Conference on Inventive Computation Technologies (ICICT)*, vol. 3, pp. 1–6 (2016).
26. Zhong, L., Wan, W., Kong, D. Javaweb login authentication based on improved MD5 algorithm. In: *IEEE International Conference on Audio, Language and Image Processing (ICALIP)*, vol. 3, pp. 131–135 (2016).
27. Z. Zheng, S. Xie, H. Dai, X. Chen and H. Wang, An overview of blockchain technology: Architecture, consensus, and future trends, in *Big Data (BigData Congress), 2017 IEEE International Congress on*, IEEE, 2017, 557–564.
28. D. Verma, N. Desai, A. Preece and I. J. Taylor, A blockchain based architecture for asset management in coalition operations, SPIE, 2017.
29. L. A. Linn and M. B. Koo, Blockchain for health data and its potential use in health IT and health care related research, in *ONC/NIST Use of Blockchain for Healthcare and Research Workshop*. Gaithersburg, Maryland, United States: ONC/NIST, 2016.
30. M. Ghorbanian, S.H. Dolatabadi, P. Siano, I. Kouveliotis-Lysikatos, N.D. Hatziargyriou, Methods for flexible management of blockchain-based cryptocurrencies in electricity markets and smart grids, *IEEE Trans. Smart Grid* 11 (5) (2020) 4227–4235 IEEE.

7

Explainable Artificial Intelligence for Cybersecurity

P. Sharon Femi[1], K. Ashwini[2*], A. Kala[1] and V. Rajalakshmi[1]

[1]Sri Venkateswara College of Engineering, Sriperumbudur, Tamil Nadu, India
[2]Amrita School of Computing, Amrita Vishwa Vidyapeetham, Chennai, India

Abstract

Artificial Intelligence incorporates human intelligence into machines and builds statistical models for providing solutions in various domains like cybersecurity. Though the AI-based techniques for detecting cyberattacks and threats are more efficient than traditional cybersecurity techniques, they lack explainability. These methods provide black-box solutions that are not transparent and interpretable in understanding the steps involved in reaching specific predictions. This reduces the confidence of users in the models used for cybersecurity, particularly in the present scenarios where cyberattacks are becoming increasingly diverse and complicated. To overcome this, Explainable Artificial Intelligence (XAI) is applied, which eventually replaces the traditional artificial, machine learning, and deep learning algorithms that operate as a black box. Given that cyberattacks are increasing day by day and the traditional AI algorithms are not sufficient for providing security, it is essential to focus more on XAI for exploiting the AI algorithms. This chapter provides a detailed discussion about explainable artificial intelligence and how it is applied in providing cybersecurity.

Keywords: Explainability, XAI, cybersecurity, cyberattacks

---

*Corresponding author: k_ashwini@ch.amrita.edu

7.1 Introduction

In recent days, more and more network attacks and mechanisms to steal data, damage reputations, hinder work and gain material advantage are being witnessed. Cybersecurity is the practice of defending and restoring networks and information systems from violations of fundamental security requirements such as data confidentiality, integrity, availability, and authenticity [9]. As the Internet becomes an indispensable tool in day-to-day life, the number of networked systems continues to grow. Also, the advances in computer networks and mobile devices have considerably increased Internet usage. This widespread use of the Internet has also lured cyber attackers to build more refined and potent cyberattack methods to their advantage. Various tools and techniques related to cybersecurity [6, 7] are designed to combat threats that target the networked systems and applications present in any organization.

To promise the confidentiality, accessibility and integrity of information transmitted on the Internet, a well-known security system must be established. Such systems are implemented to prevent financial extortion by users or reputable organizations that hinder normal business operations. Therefore, it is absolutely necessary to adopt intelligent, effective and efficient countermeasures. In addition, traditional cyber defense mechanisms are challenged by the ever-increasing amount of information circulating on the Internet [14]. Cyber hackers, on the other hand, have been striving to stay ahead of law enforcement by developing new, intelligent and sophisticated attack techniques and implementing technological advancements including AI [37]. As a result, cybersecurity researchers have started to explore AI-based approaches to improve performance.

7.1.1 Use of AI in Cybersecurity

AI techniques have delivered impressive performance on benchmark datasets in a range of cybersecurity applications like intrusion detection, fraud detection, malicious application identification, etc. [8]. Recently, machine learning–based systems outperform humans in multiple domains, including defending cyberspace. Machine learning algorithms are used in detecting anomalies and threats related to security threats and vulnerabilities [4]. Modern information defense systems and cyber systems integrate ML methods for detecting attacks and preventing negative costs.

7.1.2 Limitations of AI

Generally, AI-based security systems produce false-negative and false-positive results. The former can lead to wrong decisions, while the latter can lead to false alarms [24]. To deal with these situations, certain improvements should be made to keep the decisions explainable and reasonable [12]. If the system fails to understand and learn from cybersecurity attacks, then cybersecurity becomes a black box with pervasive quadratic risk, irrespective of how powerful and accurate the AI-based system is [36]. This growing adoption of intelligent black-box systems in high-risk environments is severely hampered by the need for transparency. This becomes a bigger problem when the machine learning models become more complex. The interpretability of machine learning models is critical for data scientists, researchers, and developers to understand the models, their value, and the accuracy of their results [26]. Thus, interpretation is required to investigate false positives, identify systematic deviations and errors, and ultimately make informed decisions for future improvements.

7.1.3 Motivation to Integrate XAI to Cybersecurity

In the above-mentioned limitations of AI-based approaches, the nature of the black box is a serious issue. This nature of AI makes the decisions of the cybersecurity systems too complex for people to realize how the output is generated. Hence, to trust the decisions of cybersecurity systems, AI must be transparent and explainable. To meet this type of requirement, several strategies have been proposed to make AI decisions more understandable by humans. This explainable technology is referred as Explainable Artificial Intelligence (XAI). XAI works by making the results produced by AI-based statistical models interpretable and enabling researchers and experts to understand causal reasoning and primary data evidence [20]. XAI provides the experts with the logical understandability of the data and the results obtained. The main motivation in integrating XAI to cybersecurity is to develop trust and to improve transparency, understandability and justifiability.

In healthcare, the implementation of XAI enables the machines to analyze data and come up with meaningful results. Second, it allows physicians to obtain decision-line information that explains how a particular decision was made.

7.1.4 Contributions

This book chapter rationalizes the motivation for integrating XAI in AI-based cybersecurity models and provides a comprehensive review of state-of-the-art XAI applications in the cybersecurity area. This chapter extensively discusses the following topics:

    i. Various forms of cyberattacks
   ii. XAI and its categorization
  iii. Frameworks for the XAI-based cyber defense mechanism
  iv. Applications of XAI in cybersecurity
   v. Challenges of XAI applications in cybersecurity
  vi. Future research directions

7.2 Cyberattacks

Cyberattacks have become more sophisticated as our society has evolved and become more interconnected. As data breaches become more common, it is critical to have a thorough understanding of modern cyberattacks. Zhang *et al.* [37] have clearly elaborated on the various cyberattacks. The XAI-based defensive solutions for various types of cyberattacks are discussed in this subsection. Figure 7.1 depicts the various forms of cyberattacks.

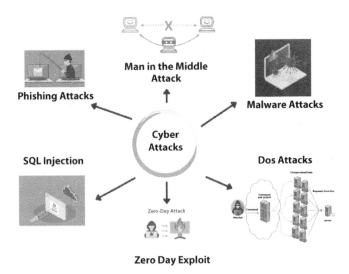

Figure 7.1 Various forms of cyberattack.

7.2.1 Phishing Attack

When a malevolent actor sends messages posing as a reliable source, it is called a phishing attack. The intention is to steal private data from the victim's computer, including login and credit card information. Phishing is an online fraud technique that is gaining recognition and takes many forms.

7.2.1.1 Spear Phishing

A scam known as "spear phishing" involves an email or other electronic contact in which the attacker tries to take money from a person, a company, or an organization. Attackers install malicious software on the victim's PC in order to access the victim's data.

7.2.1.2 Whaling

Whaling is the term used to describe a cyberattack that targets the CEO or CFO of a target organization. Attackers directly target senior or other important people by impersonating senior or other essential members of an organization. It aims to filch the money or secret data for gaining access to the computer for committing cybercrimes.

7.2.1.3 Smishing

Phishing with text messages is referred to as "SMS phishing" or "smishing." Text messages that seem to be from a reliable source are constantly sent to people who have been the target of a smishing attack. One sort of phishing attempt that occurs over the phone is voice phishing. Phishing scammers are increasingly using Voice over IP technology to communicate with their victims. A smishing effort posing as the US Postal Service was discovered by Tripwire. Malicious SMS message senders directed their victims to click on a link to learn more about an upcoming US Postal Service delivery. The malicious link took users to a number of websites designed with the sole purpose of stealing their Google account information.

7.2.1.4 Pharming

Using the social engineering technique known as "pharming," attackers redirect website users who are trying to access a specific website to a phoney one. Through unlawful websites, malware can be placed on the victim's computer to steal login credentials and personal information. Often, cybercriminals target financial organizations.

7.2.2 Man-in-the-Middle (MITM) Attack

This attack happens when a person intrudes in a conversation and impersonates one or the other, making it resemble a normal data exchange.

7.2.2.1 ARP Spoofing

ARP spoofing, sometimes referred to as ARP poisoning, is a type of MITM attack that enables attackers to eavesdrop on network device communication. By using the weaknesses in the protocol, it has the potential to poison the MAC of other devices to IP mappings using ARP. Using easily accessible tools, a malicious attacker can contaminate the ARP caches of other computers on a local network, filling them with false information.

7.2.2.2 DNS Spoofing

DNS cache poisoning or DNS spoofing is an extremely cunning cyberattack that involves poisoning the DNS cache in order to route web traffic to phishing websites. By constructing false websites that look like the user's intended destination, hackers can easily trick people into providing personal information.

7.2.2.3 HTTPS Spoofing

In HTTPS spoofing, the URL of the HTTP site of the attacker significantly differs from the URL of a genuine, legitimate site. By establishing a slightly different URL that resembles the user's intended URL but is actually slightly different, hackers are able to obtain personal information.

7.2.2.4 Wi-Fi Eavesdropping

It is the practice of hackers eavesdropping on wireless communications on unprotected networks or setting up networks with catchy names to lure users into connecting so they may steal the login information they send over that network.

7.2.2.5 Session Hijacking

This occurs when a person logs into a web page and the attacker waits for them to do so, then takes their session cookie and uses it to enter the same account from his browser. This attack is also known as a cookie session.

7.2.3 Malware Attack

Malware is a term used to describe malicious software such as ransomware, spyware, viruses, and worms. When a person accidentally clicks on a malicious link or opens an attachment of email, the machine may be infected with malware. It can restrict access to critical network resources and has the ability to gather and transmit sensitive information without the user's knowledge. Malware has the potential to disable a large number of system components, rendering the machine completely inoperable [33].

7.2.3.1 Ransomware

Ransomware is a form of malware that requires payment and is regarded as one of the most dangerous varieties. Data is encrypted, and decrypting it costs money. People unwittingly infect their computers with this type of virus through email attachments or links from dubious websites, which is one of the most common causes of infection. If ransomware is installed, it might provide hackers access to a device's backdoor, enabling them to encrypt the data on the target device and prevent its owner from decrypting it until they are paid a ransom. Because it demands ransom payments in digital currency, it is also known as crypto-malware. In conclusion, ransomware can commandeer machines, encrypt data, and ruin the victim's finances [31].

7.2.3.2 Spyware

Malware that accesses computers without the owner's consent includes spyware. This is typically done with the intention of collecting user credentials, spying on Internet activity, or acquiring private information that could be used fraudulently. The term "spyware" covers a wide range of undesirable programmes, including adware, Trojan malware, and even cookie trackers. The term "keylogger" refers to a sort of espionage software that is among the most often used. It records each keystroke performed on the keyboard and stores the data a person enters. In conclusion, spyware may invade users' privacy, gather sensitive information, steal data, or steal their identity.

7.2.3.3 Botnet

A "spider" programme known as a botnet scans the internet for security holes and stops exploitation. In this attack, the intrusion happens automatically. By inserting malicious malware into devices, it causes infection. They can be used to successfully hack into devices. They have the ability to perform distributed denial-of-service (DDoS) assaults and capture activities like keystrokes, camera images, or screenshots. Hackers utilize a botnet to gain remote access to a computer.

7.2.3.4 Fileless Malware

The software, programmes, and protocols that are inherent to or built into an operating system of a device are used to install and run fileless malware. This is memory-based and does not require downloading. It keeps wreaking havoc as long as legitimate programmes are running. Due to its stealthiness, it is hard to detect. Therefore, fileless malware has the ability to interfere with antivirus programmes and steal data.

7.2.4 Denial-of-Service Attack

DoS assaults are venomous, targeted offences that can flood a network with fraudulent requests in an effort to sabotage corporate operations. In this case, users are barred from resources that are kept on a system or in a network. They don't cause data loss.

7.2.5 Zero-Day Exploit

Zero-day security flaws are those that can be exploited by hackers. Zero-day vulnerabilities are those that have just recently been found by a user or developer. A zero-day attack happens when hackers take advantage of vulnerability before engineers can fix it.

7.2.6 SQL Injection

Malicious SQL code is used to manipulate databases on the back-end and gain access to data that is not intended for display. SQL injection-SQLI is the name of this harmful code injection method. Details like private customer information or critical business information may be jeopardized.

7.3 XAI and Its Categorization

Van *et al.* [35] coined the term XAI to depict the ability of the technology to interpret the behavior of AI-driven entities in gaming applications. The goal of XAI is to make it easier for end users to understand the results of AI. According to DARPA, XAI's goal is to generate more explainable models and enable stakeholders to better understand and appropriately trust a new generation of AI partners [31].

The research in AI has shifted to building models and algorithms that emphasize predictive power. Researchers and practitioners have recently started paying attention to XAI. XAI is a set of techniques and methods that help researchers understand and rely on the results and conclusions of machine learning models. XAI helps researchers understand the accuracy, rationality, transparency, and effectiveness of AI-assisted decision-making by comparing it to other decisions. The output needs to be interpretable in order to be credibly adjusted.

The terms explainability, interpretability, transparency and intelligibility are used to characterize XAI; the relation between these terms is depicted in Figure 7.2.

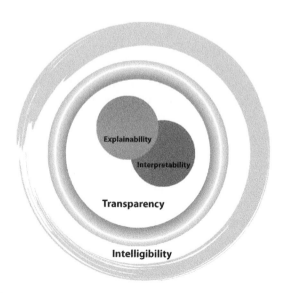

Figure 7.2 Relation between the terms of XAI.

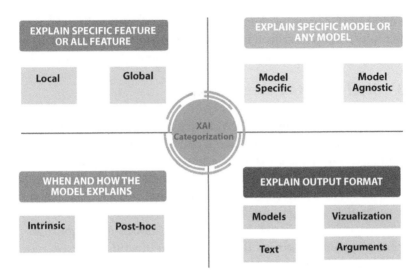

Figure 7.3 XAI categorizations.

There are numerous ways to structure XAI categories. It is clear that the some of the categorization techniques may overlap and that a particular XAI methodology may fall under one or more categories. As a result, XAI is categorized using many categorization perspectives as shown in Figure 7.3. This gives additional details and XAI approach traits at various levels.

7.3.1 Intrinsic or Post-Hoc

This categorization approach differentiates whether explainability is accomplished by restricting the complexity of the AI model (intrinsic) or by examining the methodology of the model after training (post-hoc). Using the data generated by the prediction model, an intrinsic XAI technique generates the explanation along with the prediction. Due to their inherent self-explanatory nature, some ML models, such as Decision Trees and Sparse Linear models, are recognized as intrinsic XAI techniques. Post-hoc explanations, on the other hand, include using interpretation techniques after the models have been trained and the judgments have been made. Typical post-hoc explanation techniques working independently as an external interpretable model include LIME and Permutation Importance.

7.3.2 Model-Specific or Model-Agnostic

XAI methods are categorized based on the models to which XAI is applied—model-specific or model-agnostic. Model-specific tools are unique to a particular model or set of models. For example, graph neural network explainer is a technique for delivering explanation for graph-based ML problems that are based on GNN. On the other hand, model-agnostic explanation tools can be used with any ML model. Additionally, model-agnostic explanation techniques often analyse feature inputs and outputs rather than the internal data of the models, such as weights or structural information. Tools for model-independent explanations include Grad-CAM, Saliency Map, and SHAP tools.

7.3.3 Local or Global

Depending on the scope of the decision model, explanations can be local or global. The ability of a system to explain to a user why a particular option or decision was taken is known as local explainability. This group includes certain well-liked explainability techniques including LIME [25], SHAP [19], and counterfactual justifications. Global explainability, in contrast, relates to the explanation of the learning algorithm as a whole, taking into account the training data used, the algorithms' suitable uses, and any warnings indicating the algorithm's shortcomings and improper applications. GAM is proposed as a method of global explanation to explain the distribution of neural network predictions across subpopulations.

7.3.4 Explanation Output

The format of the explanation output would have a significant impact on some users, making it another essential element of XAI categorization. For instance, text-based explanation techniques are frequently used to fine-grained information and produce comprehensible explanations in Natural Language Processing (NLP). On the other hand, the techniques to visualize explanation are employed in a wider range of fields, such as neural networks, NLP and healthcare. In reality, most feature summary statistics can also be visualized and some feature summaries can only be understood by visualization. In order to assist people to better understand the relevance of a feature, argument-based explanations require describing the features in a style that people use to make judgments. Approaches to model-based explanation must describe the internal working logic of a black-box model.

7.4 XAI Framework

As stated earlier, all the machine learning algorithms have been a black box to the users, not letting the user know what processes are carried out for a particular input and how exactly an output for a specific input is obtained. Understanding the chassis and functioning of machine learning algorithms is most important for it to be used especially in security fields. The frameworks of Explainable artificial intelligence help us to create a consecution of machine learning techniques through which more explainable models with higher performance that enable the users to understand the underlying concepts and techniques of the black box are obtained [29]. Figure 7.4 summarizes a few of the most well-known XAI frameworks, which are also detailed in this section.

7.4.1 SHAP (SHAPley Additive Explanations) and SHAPley Values

SHAP [19] is the most famous visualization tool that aids in providing detailed explanation of prediction models with which the contribution of each predictor to the final output can be examined. It can be used to simple ML algorithms like linear regression, logistic regression, decision tress and also more complex deep learning architectures that are used for image classification, natural language processing, etc. Many variants of SHAP are available. Figure 7.5 shows some of the evolution of SHAP models that have been designed based on their performance on different machine learning algorithms.

The most important component in SHAP is the SHAPley values [13]. The basic idea of SHAPley value concept has been inspired from game

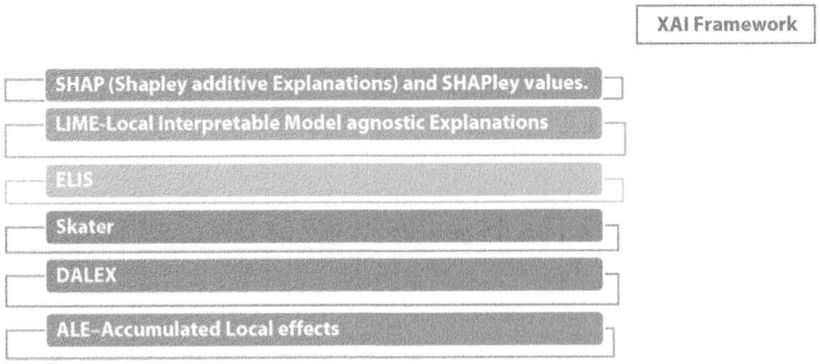

Figure 7.4 XAI framework.

theory, where each player's contribution towards the end result of the game is computed. Similarly in the SHAP architecture, SHAPley values will let us know how accurately the contribution of different features are distributed among the available features in a prediction model. The working is based on the assumption that each feature in the model works together with other features in bringing up the output. SHAPley value is mathematically defined as follows:

$$\Theta_i = \sum_{S \subseteq K \setminus \{i\}} \frac{|S|!(|K|-|S|-1)!}{|K|!} (f(S \cup \{i\}) - f(s)) \qquad (7.1)$$

where K=Set of all features,
S=Subset of features the model uses,
$f(S)$=produces predicted output for any features

7.4.1.1 Computing SHAPley Values

A brief description on the working of SHAP [20] for understanding the underlying black box algorithm is listed below.

1. From the given data, set of all possible feature combinations S are selected. They are called coalitions.
2. Average model prediction is computed.
3. Calculate the variation between the average prediction and the prediction made by the model without feature i for each coalition.
4. Calculate the variation between the average prediction and the prediction made by the model with feature i for each coalition.
5. Compute the difference for the values obtained in step 3 and step 4. This will be the marginal contribution of a feature i.
6. Average of all the values computed in step 5 gives the SHAPley values.

Figure 7.5 illustrates the evolution of SHAP models. Some of the significant flaws in SHAPley values–based architectures is that these models provide additive contribution to the explanatory variables. Thus if the model to be explored is of non-additive nature, then the SHAPley values may be misleading. Also the computation of SHAPley values is quite time consuming. However, subsampling methods can be used to address these issues.

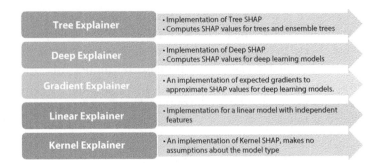

Figure 7.5 Evolution of SHAP models.

7.4.2 LIME - Local Interpretable Model Agnostic Explanations

As the name suggests, LIME [22, 34] has a property of model agnosticism, with which it treats any supervised model as a separate black box and provides explanation for it. And the explanation provided are Local explanations, i.e., explanations are provided using the samples that are in the vicinity of the observation/sample that is being explained. Working principle of LIME is almost similar to that of SHAP, with the major difference being the execution time. LIME is best suited but not limited to predicting tabular data, image and test classifiers. They are considered as a concrete implementation of intrinsic interpretable models that are trained to approximate the underlying black box model's predictions.

Given a sample test and a prediction model, the two main steps LIME does are sampling the given data to get a surrogate dataset and selecting features from the surrogate dataset that has been created in the previous step. The weights of each row of the surrogate dataset are then determined by computing how closely they resemble the original data.

Mathematically, LIME is described as:

$$Detail(x) = \arg\min_{d \in D} \{L(c, d, \Pi_x) + \Omega(d)\} \qquad (7.2)$$

Where d= explanation model for a particular instance (x)
D= All possible explanations
L= Loss function computed between explanation and prediction
c= actual black box model

$\{\Pi_x\}$ = Proximity measure,
$\{\Omega(d)\}$ = Omega complexity of the explanation model.

7.4.2.1 Working of LIME

Below is a basic explanation of how LIME operates so that one may comprehend the algorithm's black box foundation.

1. For a given set of data, discompose the data n times to create replicated feature data. This discomposed data will be with slight modification of the actual data. This data will be used by LIME to create a local linear model.
2. For the discomposed data, predict the outcome.
3. Compute the distance between each discomposed data and the actual data.
4. Distance computed is converted into a similarity score.
5. From the discomposed data, some features which best describe the predictors are selected.
6. Fit a simple basic model to the discomposed data for the selected features.
7. Coefficients of this simple model will be the explanations of the observations.

LIME in general has various proposals for dealing with explanatory variables and hence this leads to different implementations of LIME, and as a result, different results are possible [25]. Also sometimes the model may be misleading, failing to control the quality of local outfit to the data.

Another important issue to be addressed in LIME is that, with higher dimensional data, most of the data are sparse. It could be difficult to precisely define the "local neighborhood" of the relevant occurrence. Even a slight change in the neighborhood will strongly affect the explanations.

7.4.3 ELI5

ELI5 [18] is the acronym of Explain like I'm 5. It is a very famous Python package that is used to understand many ML algorithms. They are mostly used in sklearn regression and classification problems, Keras, CatBoost, etc. Various inbuilt functions can be used to get the details on how a particular decision is made in any classification or regression problems. ELI5 computes the weights for each feature and shows the contribution of each feature in predicting or classifying the output.

7.4.4 Skater

Skater [11] is another open-source library of Python for learning about the black box of any learned model. The library is capable of producing both local inference, pertaining to individual features as well as global inference, pertaining to the entire dataset.

There are many algorithms supported by Skater. Depending upon their scope of interpretation, algorithms are broadly classified as shown in Figure 7.6.

7.4.5 DALEX

Similar to the previous models, DALEX [2] is another Python library developed for Explanatory Model Analysis and can be used for both classification and regression in ML. The DALEX package adds an abstraction layer to models, enabling interaction with various models in a consistent manner.

DALEX comes with various packages with which a relationship between the model input and the model output can be easily obtained.

One of the important packages in DALEX is DALEXTRA. This package consists of various tools which aid in inspecting and improving the models. Two main functionalities DALEXTRA provides are listed and briefed below.

1. Champion-Challenger analysis
   This functionality of DALEXTRA helps us to compare two or more machine-learning models, decide which is superior, and then enhance both of them.

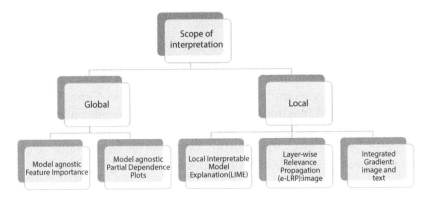

Figure 7.6 Skater algorithms.

2. Cross-language comparison
   With this functionality, explainers can be created for models that were developed in different languages.

7.5 Applications of XAI in Cybersecurity

Machine learning prediction models are trained automatically without the knowledge of the domain in which they are applied. Hence, they are known as black boxes. This ambiguity in the model-building process creates many unpredictable risks. For example, model performance drops due to out-of-domain issues that lead to poor performance, data drift, or behavior learned from historical data was unfair. There are many situations arising in various domains where the black box fails which leads to the increased interest in XAI methods.

As the frequency of devastating cyberattacks has increased, establishing, and enhancing cybersecurity is a massive social challenge. To address this social issue, a timely, most prominent and actionable intelligence [5] against the threats is developed to enable effective decisions against the cyberattacks. In this section, various applications of Explainable AI to prevent cyberattacks and techniques to provide cybersecurity are discussed. Some of the applications of XAI in cybersecurity are shown in Figure 7.7.

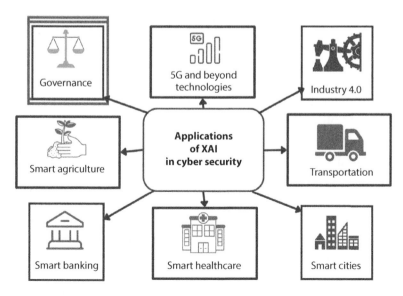

Figure 7.7 XAI in cybersecurity applications.

7.5.1 Smart Healthcare

The research carried out in smart health using artificial intelligence and machine learning is to analyze the physiological data to identify the mental health of individuals. The AI technology Human-in-the-loop system enables us to identify mental illness and give proper treatment. Early detection of mental illness would substantially reduce the damage caused by it. There are many malicious threats that target an attack against these Human-in-the-loop systems. XAI could aid in providing solutions to the cybersecurity issues to develop robust and secure Human-in-the-loop systems to handle mental illness.

A key challenge for XAI systems is to maintain the integrity of the decision-making process. To achieve this, the personnel have to take direct control of the process. The attacker can insert malicious input and train the model to provide an adversarial solution. XAI has provided efficient solutions to analyze and identify the attacks and provide cybersecurity solutions [32].

7.5.2 Smart Banking

Financial cybercrimes have been growing in recent years. In order to fortify cybersecurity in financial sectors, it is mandatory to reduce the risk score involved in digital transformation on cloud environments. This provides the banking and various financial sectors an unparalleled agility and protection to reduce cyber risk. By integrating information security into the infrastructure, data and assets are protected. This helps to establish and stabilize financial regulations and compliance program activities.

XAI cybersecurity solutions incorporate necessary policies and controls to decrease the risk involved in financial sectors. XAI-oriented fraud detection and regulatory compliance tool provides layered cybersecurity to resolve the cyberattacks [1].

7.5.3 Smart Cities

In the smart city domain, to enforce the desired service, an extended network of sensors is connected to extract data from various locations. Hence, the network infrastructure should consist of secure and reliable interconnected devices like actuators, sensors etc. The data from these devices are gathered, processed, and communicated to enable smart city services. The data from such a heterogeneous network has a great impact to enforce cybersecurity in the smart city implementation. Since these tiny low-end devices have a limitation [28] in their processing and storage capacity, it may not be robust to establish security features like authentication techniques or cryptographic encodings. This leads to possible cyberattacks on smart city services.

XAI provides solutions to prevent these attacks. It provides a transparent authentication mechanism on the devices like generating one-time passwords. All activities are controlled by proper permission to enable the action for the activity. The various software and firmware updates are made automatic and periodical. Logs are maintained in an encrypted format to prevent tampering [3]. The technology transfer is also made secure. Regular auditing and assessment, protection of access control and logging environment is taken care of to prevent potential cyber threats.

7.5.4 Smart Agriculture

The use of Internet of Things technologies, Information and Communication Technology and data analysis plays a major role in the implementation of smart agriculture and its operations. The objective of establishing smart agriculture technologies is to meet the growing demands to monitor crops, check soil fertility, automate soil testing whenever needed, etc. Smart agriculture depends on the smart networks forming cyber-physical systems [16]. These systems enable communication with other devices such as sensors, processing control units, etc. The entire setup is controlled by computer and communication systems. These devices collect data such as moisture, weather fluctuations, fertilization and so on. This data is further analyzed and processed to improve the productivity of agriculture. All these activities increase the cyber threats as the data is collected from various sources [27]. The XAI provides cybersecurity solutions to handle the physical attacks, attacks against authentication, replay attacks and malicious code attacks.

7.5.5 Transportation

Intelligent transportation systems face a big challenge in working with increased value of ITS data and connectivity [23]. There are many cyber risks involved in data collection and processing of data.

Some of the major impacts of cyberattacks in the transportation include:

- The most important data files and information are blocked
- Traffic lights, toll booths are disrupted
- Payroll services are interrupted

- Ticket machines and fair gates are interrupted
- Sensitive information from emails is stolen
- Personal information is stolen

The XAI-based ITS architecture employs risk-reducing components that are extensible, interoperable, and can operate in degraded conditions in times of cyberattacks.

7.5.6 Governance

Governance is a set of wise decisions made by organizations to enable cybersecurity activities. The tenets of a cybersecurity programme offers a comprehensive picture of risk and active tracking of performance [10]. Organizations can leverage the advantages of functioning in the digital market with the help of strictly delineated cybersecurity governance. The effectiveness and durability of the digital business transformation are supported by effective cybersecurity governance. Without sound cybersecurity governance, companies would struggle to keep the trust of external stakeholders or guarantee core business sustainability [21]. Implementing effective XAI for cybersecurity in governance would enable the following:

- A cybersecurity vision drives overall organizational decision-making in accordance with the overall plan.
- Supervision and resource allocation via a platform and training for cybersecurity.
- A comprehensive approach to risk assessment that takes into account cybersecurity risk and improves knowledge of the organization's exposure to cyber threats.
- Duties and responsibilities for cybersecurity that is clearly defined, allocated, and incorporated into the enterprise.
- A reliable mechanism for monitoring the implementation and commenting on progress.

7.5.7 Industry 4.0

Interconnected and smart technologies are a part of work in organizations, and even wearable devices are being used by humans. This shows the benefit of emerging technologies from artificial intelligence (AI), machine learning (ML) and robotics to quantum computing. Even the Internet of Things (IoT) and additive manufacturing benefited from the smart technologies [17]. As the Industry 4.0 is becoming smarter day by day by means of interconnected devices, the cyber threats are even increased at the other end [15].

A few of the cyber risks in the Industry 4.0 are listed below:

- Critical application and infrastructure are accessed and controlled by attackers by evading the detection caused by state adversaries.
- New sophisticated approaches by the attackers intensify the malicious attacks on the critical cloud infrastructure.
- The latest vulnerabilities are adapted by the targeted and crime actors that will exploit the trusted sources and supply chain.
- Stolen identities and credentials are exploited by sophisticated adversaries. This leads to the ransomware hunting attack.

To handle all these cyber threats, XAI applications enable us to define the solution based on the particular attack rather than a common solution for all the attacks.

7.5.8 5G and Beyond Technologies

Attacks by nation-state hackers are becoming more of a threat, especially for telecom companies, which were lately witnessed. Additionally, cutting-edge technologies like 5G pose new threats and vulnerabilities. However, it is crucial to note that 5G also offers significant advantages, particularly enhanced security features like improved authentication and encryption features [30]. The detailed requirements are currently being created, and much would depend on the way they will be implemented into products and used by operators to handle cyberattacks. In spite of its own secured infrastructure, there are cyberattacks in 5G technologies. The most prominent problem-oriented solutions are modeled using XAI on 5G cyberattacks.

7.6 Challenges of XAI Applications in Cybersecurity

A review on the XAI techniques has been done for the different attacks that have been happening and the cybersecurity domains. XAI is a powerful tool but there are also a couple of challenges [37] that they face which are discussed in detail.

7.6.1 Datasets

The main problem with datasets is that they are not updated in certain directions. This can occur due to confidentiality and ethical issues. If the cyberattacks that happened recently are integrated in the establishment of cyberattack defense mechanisms it leads to ineffectiveness of the XAI application, which is why they are not included in the datasets. Since cyberattacks are becoming more complicated, the datasets have to be updated well. The other problem that the datasets face is that there is a deficit in voluminous amount data that is needed for the XAI training methods which will result in decreased performance and explainability of the XAI approaches. The information pertaining to cyberattacks and cyber industries is redundant and not balanced. The challenge for the XAI models is the heterogeneity present in the dataset. These problems give us an outlook on the present voluminous benchmark datasets used for training and testing.

7.6.2 Evaluation

Evaluation for the XAI systems plays an important role. The performance of the cybersecurity systems includes performance metrics such as Precision, F1-Score and ROC. XAI systems must be able to assess the quality, value, and satisfaction of explanations, etc. But the challenges faced by XAI systems are more generic. The XAI explanation evaluation measurements are divided into two categories, namely, user satisfaction and computational measurements. User satisfaction-based evaluation causes privacy issues because they are independent on user feedback or interview. Inherently interpretable models are utilized by many researchers for computational measurements. They lack certain things in the other cybersecurity domains like computational resources and computational power. To provide future improvements for XAI applications it is required to consider a set of standard evaluation metrics.

7.6.3 Cyber Threats Faced by XAI Models

The XAI models are encountering many cyberattacks targeting the vulnerabilities of the explanation approaches, which makes it dangerous for the cybersecurity systems. For instance, most popular XAI explanation methods such as LIME and SHAP, deployed in the XAI application of cybersecurity, can also be fooled. The most defensive cyber approach is the security of the performance of the prediction results of the XAI models.

It is important to retain the transparency and efficiency of the entire system and also to prevent the cyberattacks.

7.6.4 Privacy and Ethical Issues

One of the crucial challenges faced by XAI models is to consider privacy concerns. Authentication, emails, and password are part of a person's right to their personal privacy. It is important to be cautious in ensuring that there is no discrimination, bias, or unfairness made by the XAI system and the explanations that go along with them. In the specific domain of cybersecurity, XAI terms can be eliminated. The privacy and security-related concerns increase as the data is collected from various security-related sources and only authorized persons are provided access to XAI models.

7.7 Future Research Directions

One potential area for future research is the creation of both high-quality and updated datasets that can be used for XAI applications for cybersecurity. Research on the trade-off between performance and explainability of XAI techniques used in cybersecurity is lacking. Future research could focus on how to develop customer-centered XAI systems for cybersecurity to enhance customer understandability and performance without compromising security. Even though current studies on cyber threats and corresponding defensive mechanisms are focusing on the performance of AI models, the adversarial threats and defences against the explainability of XAI models still need to be explored. As privacy and ethical concerns have recently received attention, confidentiality and data protection are important challenges in the field of cybersecurity. Future research may focus on the XAI systems' generated explanations and data protection.

7.8 Conclusion

This chapter discusses the key insights regarding using XAI for cybersecurity. With the use of ML models and the XAI framework, predictions may be understood and interpreted. An application of AI called cybersecurity analyses datasets and keeps track of a variety of security vulnerabilities and fraudulent activity. The work that is being presented offers a cutting-edge analysis of XAI in cybersecurity. The concept of cybersecurity is introduced first, stressing the many forms of cyberattacks and their effects. An XAI system boosts confidence in the XAI-based cybersecurity system by offering explanations. XAI explanations of how user data is used in algorithmic decision-making could teach end users. The visualization and explainability of the XAI system can help cybersecurity professionals assess the reliability and uncertainty of models. This was followed by a thorough analysis of the most recent XAI study findings.

References

1. Adadi, A. and Berrada, M., 2018. Peeking inside the black-box: a survey on explainable artificial intelligence (XAI). *IEEE Access, 6*, pp. 52138-52160.
2. AL-Essa, M., Andresini, G., Appice, A. and Malerba, D., 2022. XAI to Explore Robustness of Features in Adversarial Training for Cybersecurity. In *International Symposium on Methodologies for Intelligent Systems* (pp. 117-126). Springer, Cham.
3. Alibasic, A., Al Junaibi, R., Aung, Z., Woon, W.L. and Omar, M.A., 2016, September. Cybersecurity for smart cities: A brief review. In *International Workshop on Data Analytics for Renewable Energy Integration* (pp. 22-30). Springer, Cham.
4. Almseidin, M., Alzubi, M., Kovacs, S. and Alkasassbeh, M., 2017, September. Evaluation of machine learning algorithms for intrusion detection system. In *2017 IEEE 15th International Symposium on Intelligent Systems and Informatics (SISY)* (pp. 000277-000282). IEEE.
5. Arrieta, A.B., Díaz-Rodríguez, N., Del Ser, J., Bennetot, A., Tabik, S., Barbado, A., García, S., Gil-López, S., Molina, D., Benjamins, R. and Chatila, R., 2020. Explainable Artificial Intelligence (XAI): Concepts, taxonomies, opportunities and challenges toward responsible AI. *Information Fusion, 58*, pp. 82-115.
6. Ashwini K, User Name-Based Compression and Encryption of Images Using Chaotic Compressive Sensing Theory, *Computer Journal*, 2022; bxac175, https://doi.org/10.1093/comjnl/bxac175
7. Ashwini, K., & Amutha, R. (2018). Fast and secured cloud assisted recovery scheme for compressively sensed signals using new chaotic system. *Multimedia Tools and Applications, 77*(24), 31581-31606.
8. Chan, L., Morgan, I., Simon, H., Alshabanat, F., Ober, D., Gentry, J., Min, D. and Cao, R., 2019, June. Survey of AI in cybersecurity for information technology management. In *2019 IEEE technology & engineering management conference (TEMSCON)* (pp. 1-8). IEEE.
9. CISA, What is Cybersecurity? https://www.cisa.gov/uscert/ncas/tips/ST04-001 (accessed Jul. 01, 2022).
10. Dor, L.M.B. and Coglianese, C., 2021. Procurement as AI governance. *IEEE Transactions on Technology and Society, 2*(4), pp. 192-199.

11. Dwivedi, R., Dave, D., Naik, H., Singhal, S., Rana, O., Patel, P., Qian, B., Wen, Z., Shah, T., Morgan, G. and Ranjan, R., 2022. Explainable AI (XAI): core ideas, techniques and solutions. *ACM Computing Surveys (CSUR)*.
12. Gerlings, J., Shollo, A. and Constantiou, I., 2020. Reviewing the need for explainable artificial intelligence (xAI). *arXiv preprint arXiv:2012.01007*.
13. Giudici, P. and Raffinetti, E., 2022. Explainable AI methods in cyber risk management. *Quality and Reliability Engineering International*, 38(3), pp. 1318-1326.
14. Gümüşbaş, D., Yıldırım, T., Genovese, A. and Scotti, F., 2020. A comprehensive survey of databases and deep learning methods for cybersecurity and intrusion detection systems. *IEEE Systems Journal*, 15(2), pp. 1717-1731.
15. Javaid, M., Haleem, A., Singh, R.P. and Suman, R., 2022. Artificial intelligence applications for industry 4.0: A literature-based study. *Journal of Industrial Integration and Management*, 7(01), pp. 83-111.
16. Junaid, M., Shaikh, A., Hassan, M.U., Alghamdi, A., Rajab, K., Al Reshan, M.S. and Alkinani, M., 2021. Smart agriculture cloud using AI based techniques. *Energies*, 14(16), p. 5129.
17. Khan, I.H. and Javaid, M., 2021. Role of Internet of Things (IoT) in adoption of Industry 4.0. *Journal of Industrial Integration and Management*, p. 2150006.
18. Kuzlu, M., Cali, U., Sharma, V. and Güler, Ö., 2020. Gaining insight into solar photovoltaic power generation forecasting utilizing explainable artificial intelligence tools. *IEEE Access*, 8, pp. 187814-187823.
19. Lundberg, S.M. and Lee, S.I., 2017. A unified approach to interpreting model predictions. *Advances in Neural Information Processing Systems*, 30.
20. Lundberg, S. (2019). *SHAP (SHapley Additive exPlanations)*. Python package.
21. Mäntymäki, M., Minkkinen, M., Birkstedt, T. and Viljanen, M., 2022. Defining organizational AI governance. *AI and Ethics*, pp. 1-7.
22. Mishra, S., Sturm, B.L. and Dixon, S., 2017, October. Local interpretable model-agnostic explanations for music content analysis. In *ISMIR* (Vol. 53, pp. 537-543).
23. Mugurusi, G. and Oluka, P.N., 2021, September. Towards Explainable Artificial Intelligence (XAI) in Supply Chain Management: A Typology and Research Agenda. In *IFIP International Conference on Advances in Production Management Systems* (pp. 32-38). Springer, Cham.
24. Nikolskaia, K.Y. and Naumov, V.B., 2021, September. The Relationship between Cybersecurity and Artificial Intelligence. In *2021 International Conference on Quality Management, Transport and Information Security, Information Technologies (IT&QM&IS)* (pp. 94-97). IEEE.
25. Ribeiro, M.T., Singh, S. and Guestrin, C., 2016, August. " Why should i trust you?" Explaining the predictions of any classifier. In *Proceedings of the 22nd ACM SIGKDD international conference on knowledge discovery and data mining* (pp. 1135-1144).

26. Rudin, C., 2019. Stop explaining black box machine learning models for high stakes decisions and use interpretable models instead. *Nature Machine Intelligence, 1*(5), pp. 206-215.
27. Sabrina, F., Sohail, S., Farid, F., Jahan, S., Ahamed, F. and Gordon, S., 2022. An interpretable artificial intelligence based smart agriculture system. *Computers, Materials & Continua*, pp. 3777-3797.
28. Sanchez, L., Muñoz, L., Galache, J.A., Sotres, P., Santana, J.R., Gutierrez, V., Ramdhany, R., Gluhak, A., Krco, S., Theodoridis, E. and Pfisterer, D., 2014. SmartSantander: IoT experimentation over a smart city testbed. *Computer Networks, 61*, pp. 217-238.
29. Šarčević, A., Pintar, D., Vranić, M. and Krajna, A., 2022. Cybersecurity Knowledge Extraction Using XAI. *Applied Sciences, 12*(17), p. 8669.
30. Shokoor, F., Shafik, W. and Matinkhah, S.M., 2022. Overview of 5G & Beyond Security. *EAI Endorsed Transactions on Internet of Things, 8*(30).
31. Srivastava, G., Jhaveri, R.H., Bhattacharya, S., Pandya, S., Maddikunta, P.K.R., Yenduri, G., Hall, J.G., Alazab, M. and Gadekallu, T.R., 2022. XAI for Cybersecurity: State of the Art, Challenges, Open Issues and Future Directions. *arXiv preprint arXiv:2206.03585*.
32. Srinivasu, P.N., Sandhya, N., Jhaveri, R.H. and Raut, R., 2022. From Blackbox to Explainable AI in Healthcare: Existing Tools and Case Studies. *Mobile Information Systems, 2022*.
33. Sun, R., Botacin, M., Sapountzis, N., Yuan, X., Bishop, M., Porter, D.E., Li, X., Gregio, A. and Oliveira, D., 2020. A praise for defensive programming: Leveraging uncertainty for effective malware mitigation. *IEEE Transactions on Dependable and Secure Computing*.
34. Staniak, M., Biecek, P., Igras, K., and Gosiewska, A. (2019). *localModel: LIME-Based Explanations with Interpretable Inputs Based on Ceteris Paribus Profiles*. R package version 0.3.11.
35. Van Lent, M., Fisher, W. and Mancuso, M., 2004, July. An explainable artificial intelligence system for small-unit tactical behavior. In *Proceedings of the national conference on artificial intelligence* (pp. 900-907). Menlo Park, CA; Cambridge, MA; London; AAAI Press; MIT Press; 1999.
36. Zeadally, S., Adi, E., Baig, Z. and Khan, I.A., 2020. Harnessing artificial intelligence capabilities to improve cybersecurity. *IEEE Access, 8*, pp. 23817-23837.
37. Zhang Z, Hamadi HA, Damiani E, Yeun CY, Taher F. Explainable Artificial Intelligence Applications in Cyber Security: State-of-the-Art in Research. arXiv preprint arXiv:2208.14937. 2022 Aug 31.

8

AI-Enabled Threat Detection and Security Analysis

A. Saran Kumar[1]*, S. Priyanka[2], V. Praveen[1] and G. Sivapriya[2]

[1]Bannari Amman Institute of Technology, Sathyamangalam, Tamil Nadu, India
[2]Kongu Engineering College, Erode, Tamil Nadu, India

Abstract

Threat detection is a way of analyzing the entire system to predict the malicious things that take place over the network. Generally, threat detection methods include configuration, indicator, modelling and threat behavior. Currently, a Phishing attack is one of the popular attacks that happen in the internet and it grows in an exponential manner. It steals highly sensitive information through a website that closely resembles an authorized website. Phishing includes not only sending emails and waiting for the reply but also includes taking the information that is bypassed through digital communication medium. Normally technophile will induce the user to collect the information from managerial assets and networks. Recently, Artificial intelligence is the best way to analyze the vast amount of data. With the advent of artificial intelligence techniques, threat detection software behaves like a technophile which in turn helps to identify the cyber criminals. Currently, several Deep Learning algorithm can be used to predict phishing websites and anomalous behavior. This work incorporates recurrent neural networks combined with Adam optimizer to build a hybrid learning model to assess whether a website URL is good or bad. The proposed model outperforms the existing various deep learning models with accuracy of 97%, precision of 97%, recall of 98%. F1 Score is 97%.

Keywords: Hacking, threat detection, security analysis, phishing attack, feature extraction, multilayer perceptron, Adam optimizer, recurrent neural networks

---

*Corresponding author: sarankumar@bitsathy.ac.in

S. Sountharrajan, R. Maheswar, Geetanjali Rathee, and M. Akila (eds.) *Wireless Communication for Cybersecurity*, (175–198) © 2023 Scrivener Publishing LLC

8.1 Introduction

The web is an important channel for organizing [1] business, accessing news, financial transactions, playing games, interacting with government bodies, entertainment, and other types of services. In this digital era, cybersecurity has become a serious issue. A cyberattack is a process of stealing another person's information or organization in a malicious manner by an individual or organization. Malware can happen through accessing network key components, installing harmful software, spyware, information coming from a reputable organization, etc. Threat detection is a way of analyzing the entire system to predict the malicious things that take place over the network. Generally, threat detection methods include configuration, indicator, modelling, and threat behavior. Nowadays, phishing is one of the serious cyber threats that spread across the country.

8.1.1 Phishing

Phishing (Figure 8.1) is an attack which involves getting sensitive information of a target related to personal details, master card details, login information, banking, etc., via email, websites, social networks, or messages. The main source of a phishing attack is through email and websites. Phishers can also introduce content into the targeted system, and they can modify the email address in a way that resembles an original email address. With the help of this information phishers can access an accounts section

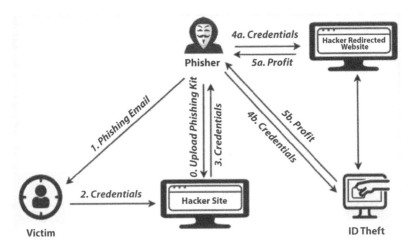

Figure 8.1 Phishing attack.

and that results in monetary loss [2]. Apart from phishing, voice phishing is also available, and several types of phishing strategies are continuously developed by cybercriminals. Increased usage of social media affords a fertile field for phishing assaults because of growing sharing of personal details. Statistics say that 250 million apps are downloaded per day.

Attacks are classified into three types [3]:

- Attack initiation
- Data collection
- System penetration

 i. Attack initiation: It involves two categories, technical and behavioral attack. The first involves penetrating suspicious email into spoofed mail. The second attack will be focused on getting their sensitive information.
 ii. Data collection: It is about collecting the information from targets that occurred during interaction with the materials of attack. This can be done automatically or in a manual manner. Automated data collection is mainly based on forged web forms, recorded messages, key loggers, recorded messages, event invitation, awarding rewards, and so on.
 iii. System penetration: It makes use of resources of the system to make the initiation of the phishing attack easier. Fast-flux and cross-site scripting are the two strategies for penetration.

A phishing attack [4] happens mainly on a personal computer system due to the following five reasons: Clients don't possess short data about Uniform Resource Locator (URLs); the specific thought regarding which pages can be relied upon; whole area of the page in light of the redirection or secret URLs; the URL has numerous potential choices; or a few pages are unintentionally entered and users can't separate a phishing site page from the genuine pages. Figure 8.2 represents the statistics of a phishing attack from a 2021-2022 report (https://docs.apwg.org//reports/apwg_trends_report_q1_2022.pdf).

Timely and effective phishing detection of URLs is critical [5]. It successfully safeguards the internet users from a phishing attack. On the client-side web browser, a blacklist is offered by certain services like Microsoft Smart Screen Filter and Google Safe Browsing, depending upon surveys collected by URLs directly and their corresponding pages. On the side of

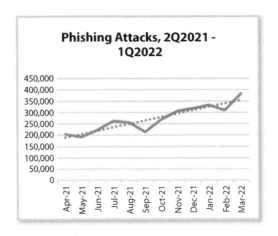

Figure 8.2 Phishing attack survey from 2021-2022.

servers, phishing detection renovate their phishing blacklist URLs based on experts of security scan, endorsements from advanced clients or experienced researchers are gathered with the URLs and respective web pages. During detection of phishing, a false positive has the potential to oppose the user's identity to visit a legitimate website whereas a false negative will make victim as user for the phishing attack. Based on blacklisting, detection of phishing can attain an approximately zero false positive rate. However, it is not possible to add new phishing URLs to the blacklist in timely fashion, and many false negatives can happen in repetition. In contrast, experience-based fraud identification can lead to some false positives, but it has the advantage of identifying real-time current phishing URL.

8.1.2 Features

- Eye-catching statements
- Sense of extremity
- Attachments
- Hyperlinks
- Unusual sender
- Website replica

Many researchers have proved that detecting a phishing attack with machine learning algorithm using heuristics approaches can attain higher accuracy with less false positive rates. They used two kinds of URL features—lexical and host features. Lexical features are features of the text, such as word and n-gram statistics of the URL string itself, and host features like domain, registration, and URL's hosting server location features.

To protect against this attack, an automated mechanism is required to detect this malicious content at an earlier stage before it gets to the user. One of the main advantages of using a deep learning model here is because there is no need to do feature extraction manually. Recurrent neural network (RNN) is used to predict the phishing websites. A neural network is a machine learning model comprising artificial neurons, and they are interconnected. RNN is one of the artificial neural networks where it is suitable to model sequential patterns. The unique characteristics of RNN is the time to the model and it helps to permit them to process data sequentially one at a time and grasp their sequential dependencies. An RNN model will consider input from the previous step outcome. This remembering of the previous step can be done with the help of a hidden layer in recurrent neural network model, and it has a memory which retains few information about the sequence. Also, RNN gives a high prediction rate due to the presence of more hidden layers in its architectural design.

8.1.3 Optimizer Types

Optimizer is a module which updates the neural network attributes like weight and learning rates. It is used to increase the production and decrease the error rate as shown in Figure 8.3 below. There are different types of optimizers which are Gradient Descent (GD), Mini-Batch GD, Stochastic GD, SGD-Momentum, Adadelta, Root mean square (RMS) prop, Adagrad and Adam optimizer are utilized to reduce the loss function.

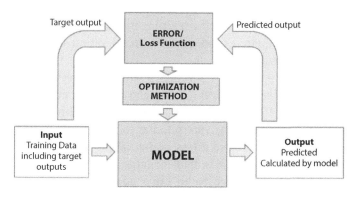

Figure 8.3 Optimizer workflow.

8.1.4 Gradient Descent

Gradient Descent optimizer is the most traditional optimizer used for solving the case of convex optimization problem. To predict the cost function gradient to the parameters needs large memory and decrease the process since gradient is estimated for the entire dataset in one epoch.

Stochastic Gradient Descent: It inherits the properties of gradient descent. This optimizer deals with a non-convex optimization issue. Instead of going batch processing it works on a single update at a time. To determine all the parameter value, it used same learning rate. It is rarely used in application because of the slow speed of computation, and frequent updates are expensive.

Mini-Batch Gradient Descent combines SGD and batch gradient optimizer. Training data were divided into batches and updates are performed in a batch-wise manner. It does not make a promise to provide good convergence for all times and it occupies less memory.

SGD with Momentum deals with adding momentum to regular Stochastic Gradient Descent. Using the momentum leads to a reduction of the noise ratio; incorporating extra hyperparameter is one of the drawbacks for this optimizer.

Adagrad optimizer is adaptive gradient optimization algorithm. For determining the upgraded parameter value, the learning rate plays an essential role. For each epoch, various learning parameters are being used by this optimizer and it is very good in nature for sparse data. Adadelta is an addition of adaptive gradient optimizer, and it is responsible to take charge of violent nature of decreasing the learning rate microscopically.

Adadelta and Root Mean Square Propagation optimizing algorithms evolved at the same time to sort out the destructive learning rate of adaptive gradient problem. Two of them used Exponential Weighted Average to find out the learning rate. Root Mean Square Propagation is an adaptive learning technique which splits the learning rate through an exponentially weighted mean of squared gradients. For accelerating the optimization process, RMS prop will be used.

Adam optimizer is Adaptive Moment Estimation method to compute learning rate adaptively for every parameter for each epoch. It utilizes the combination of RMS prop and gradient descent with momentum to find the parameter values. It is very popular for non-convex optimization problem and memory required is less, computation time is faster, it is best for the mobile objectives, and suited for large data and parameters with good computation. Compared to all other adaptive learning algorithms optimizer Adam sounds good to train the neural network in minimum time and gives the best result.

8.1.5 Types of Phishing Attack Detection

Phishing detection attacks can be classified into three types: heuristic and machine learning–based approach, proactive phishing approach, and phishing-based black list and white list approaches. In heuristic and machine learning approach, class labels are available in the dataset in order to make the prediction process correctly. It is based on supervised and unsupervised learning algorithms. Proactive phishing approaches are similar to machine learning approaches which will help users to identify the URLs as legitimate or malicious by processing the URLs' information. Black list and white list approaches are traditional methods for phishing URL detection. These methods are currently not in use due to the growth of web contents as the phishing process is tedious to predict.

8.2 Literature Survey

Anand Joseph Daniel [9] introduced a Machine Learning (ML) based algorithm for the detection of insecure websites which attracts online users to obtain their username and password. The proposed methodology used a mixture of Random Forest (RF) and Support Vector Machine (SVM) classifiers to detect and categorize the phishing websites into three types, namely malicious, spam and benign. The proposed methodology achieves an Accuracy (ACC) of 90% by testing with GENI phishing dataset. The main drawback is the non-consideration of the external factors which will reduce the ACC of the model when tested by a large-sized database.

It might be difficult to build good classification models using skewed training data. Mahmoud [10] introduced RUS Boost, a fresh approach that addresses the issue of class imbalance. When training data is unbalanced, RUS Boost combines data sampling and boosting to offer a quick and easy way to improve classification performance. Additionally, to RUS Boost is outperforming SMOTE Boost, another hybrid sampling/boosting algorithm, in terms of performance. It has significantly quicker model training times and is less computationally expensive than SMOTE Boost.

Brad Wardman [11] proposed a set of file-matching methodologies to detect the set of websites affected by a phishing attack using content-based approaches that cause intrusion to the user's content. The file-matching algorithms used include main index matching, phishDiff, deep MD5 matching and context triggered piecewise hashing algorithms. The dataset used for the research consists of 49,840 URLs for which web contents are available. The URLs are matched with the Cyveillance company dataset labels in order to check for phished URL content. Each of the string matching algorithms identifies the phished URLs by comparing with the Cyveillance company dataset. The PhishDiff gives a high prediction rate when compared with the other algorithms. The proposed approach gives an accuracy greater than 90%; the main drawback of the approach is that more time is taken for the matching process.

Mohammad Nazmul [12] proposed a framework using a set of ML algorithms, namely decision tree and RF. The proposed approach was tested on phishing dataset collected from Kaggle repository. In order to reduce the size of the dataset and to select the important attributes from the dataset for faster detection of phishing websites, principal component analysis was utilized. When performing the detection of phishing websites, random forest algorithm performs better than decision tree algorithms and gives an ACC of 97.2%. The proposed method works well with the benchmarked dataset only, which is the drawback of the approach.

Wenwu Chen [13] proposed an effective approach using Long short-term memory (LSTM) which is based on Recurrent neural network (RNN) for the detection of phishing websites. The dataset used for the work was gathered from Yahoo and Phishtank. The dataset was classified into two labels, namely original and phished. The features are reduced in order to effectively predict the phished websites by using dimensionality reduction. The proposed approach outperforms the traditional RNN by giving accuracy of 97%. The main issue with the proposed method is the large training time for the training of the neural network.

Sountharrajan [14] proposed a deep learning–based approach using Deep Boltzmann Machine (DBN) and Stacked Auto encoder (SAE) for the detection of phishing URL. The dataset used for the study was collected from Kaggle repository. Initially feature selection process was carried out for the reduction of dimensions of the dataset in order to effectively classify the phished URLs. The dataset was categorized as 80% and 20% for training and testing phase, respectively. After feature selection process, classification was done using Deep neural network (DNN). The accuracy of the proposed work is more than 85%. The main drawback of this method is the difficulty involved in training of the multiple layers of the neural network.

Fatima Salahdine [15] proposed an ML-based methodology for finding phishing websites. More than 4,000 phished emails sent to North Dakota University were collected and analyzed for the detection process.

The important features from the emails are selected using feature selection process and used for testing and training purposes. Algorithms like SVM, Logistic regression (LR) and ANN was used for the detection of phished websites. Various ACC metrics like false positive rate, true positive rate, recall and precision were considered in the study. SVM with radial basis function outperforms all other classifiers in the prediction of phished websites. The main drawback of the study is the correct selection of feature selection process for the detection of malicious websites so that privacy of the information is preserved.

Ishita Saha [16] proposed a deep learning approach using multi-layer perceptron (MLP) algorithm for the detection of malicious websites. The dataset for the study was gathered from Kaggle repository and has information about 10,000 websites. The detection method involves three phases, namely data collection, pre-processing, and classification of websites. In the pre-processing stage, feature selection methods are employed to select those subset of features used for classification. While training the model, each layers are trained in order to perform the detection correctly in the testing phase. The proposed detection model yields an accuracy of 95%. Training each layers of the neural network takes more time, which makes the model difficult to use.

Ram Basnet [17] proposed a novel framework for the detection of phishing attacks using machine learning algorithms. The dataset used for the study was collected from the Ham corpora and Phishing corpus which has a combination of both legal and illegal emails. Biased SVM and Self-organizing maps (SOMs) was used for the classification purposes. Also, clustering algorithm, namely k-means algorithm, was used to cluster the set of legal websites. Biased SVM–based approach yields an accuracy of 90% when compared with SOM algorithm. Ranking of features must be performed in order to yield more accuracy, which is the drawback of the proposed approach.

Alfredo Cuzzocrea [18] proposed an ML-based approach to detect the phishing websites using Decision tree (DT) algorithm. The PhishTank dataset was used for the study in which only 10 features from the dataset were considered for the prediction. The proposed method has a learning step and a prediction step. In the learning step, feature vector is generated for the dataset and used for the model building, whereas in the prediction step, the test dataset is fed as input to the trained model and output is analyzed to measure the performance of the proposed framework. The time taken for the learning phase is somewhat high, which makes the proposed model less likely to be used by the researchers.

Due to less data security, phishers can steal important information stored in a cloud environment. Dutta [19] proposed a neural network–based algorithm for the detection of malicious URLs that cause phishing attacks. Phishtank and AlexaRank datasets that consists of both legitimate and phished websites were used for the study. During the training phase, features are extracted which are then used to build the classification model. The model is then tested using various datasets for finding the accuracy of the proposed system. The model has found 7,900 phished websites from 10,000 input websites, which results in high accuracy of the built model.

A URL-based machine learning anti-phishing technique was suggested by Jain and Gupta [20]. To verify the effectiveness of their strategy, the authors took 14 attributes from the URL to identify the website as malicious or authentic. The suggested method was trained using nearly 33,000 phishing and legitimate URLs in Naive Bayes (NB) and SVM classifiers. The process of learning was the main emphasis of the phishing detection approach. They identified 14 distinct features that distinguish authentic websites from phishing websites. As websites with SVM classification are found, the results of their trial have over 90% accuracy.

8.3 Proposed Work

A classic example of multilayer perceptron (MLP) is the neural network (NN) model that a hierarchical collection of neurons or units for high-level computation. Due to its layered architecture, it is possible to extract large number of features from the simple data which in turn helps for easy prediction process. The versatile structure of NN makes it more suitable for feature extraction and learning process. The flow diagram of the proposed work is shown in Figure 8.4.

The dataset used for the study divided into 70% for training phase and 30% for testing phase. The input dataset is first pre-processed for the removal of inconsistent data and then important features are extracted from the data by using feature extraction algorithms in order to be used for the classification process. After the feature extraction process, the extracted features are given to Adam optimizer for fine tuning of required features. Lastly, the extracted features are given to the classifier for the prediction process. The prediction process categorizes the URL as either phished, legitimate or suspicious so that the user can easily identify the malicious websites. The phished URLs are fraudulent and can cause harmful actions when the user clicks on them. Legitimate URLs are the trusted type where

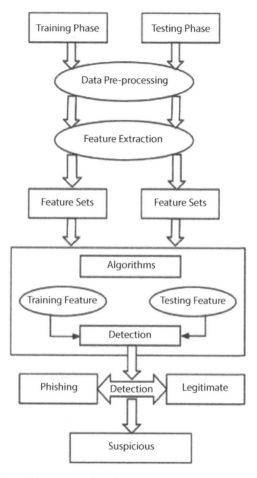

Figure 8.4 Flowchart of the proposed work.

users can use it. Suspicious URLs are similar to official URLs that cheat the users in order to cheat and harm them.

Recurrent neural networks (RNN) are highly used for the processing of sequential data which has the ability to learn deeply using its different layers of neurons for correct prediction. RNNs share the benefit of Markov chain models, in that they process data sequentially, taking into consideration the data order. Typically, the input text is reduced to a series of letters, words, or phrases. Since RNNs form the foundation for the current trending language classifier models, it can be used for exact classification of emails so that phishing can be highly detected.

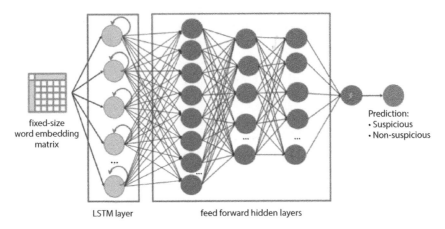

Figure 8.5 Architecture of RNN.

It is important to note that prior works have utilized RNNs to classify harmful URLs and websites, which is a concern [21]. Researchers have utilized a number of factors and accurately identified websites and URLs as phished, suspicious and legitimate. The phished website detection problem can be categorized as a binary classification problem as there are only two classes, namely ham URLs and malicious URLs. It is straightforward to abstract this classification to predicting the value of x, where x equals 1 if phish and 0 if ham. The structure of RNN is shown in Figure 8.5.

The RNN implementation can be represented mathematically (equation 8.1) as,

$$h(t) = f_H\left(W_{IH}x(t) + W_{HH}h(t-1)\right)$$
$$y(t) = f_O\left(W_{HO}h(t)\right) \quad (8.1)$$

Here $x(t)$ and $y(t)$ represents input and output vectors
W, W_i, W_o represents weight matrices of the network
f and f_o represents hidden activation function and output activation function, respectively.

8.3.1 Data Collection and Pre-Processing

The proposed model predicts the good and bad connection website Uniform Resource Locator (URLs). Web page Phishing Detection Dataset contains a wide variety of data related to intrusions and anomaly URLs. This dataset comprises 11,430 URLs with 87 extricated features. It acts as base to detect phishing URLs using machine learning algorithms. Features are categorized into three classes. From the structure and syntax of URL, 56 features are extracted. From the content of corresponding pages, 24 features were extracted and finally 7 features are extracted by querying external services. It maintains 50% phishing URLs and 50% legitimate URLs in a balanced manner. Dataset comprises URL length, host name length, IP, Path length, URL entropy, length ratio, punctuation count, suspicious word count, etc. A series of experiments was conducted to analyze the recurrent neural network–based classifier by using Web page Phishing Detection Dataset [6].

A Uniform Resource Locator is a Uniform Resource Identifier that will locate the available resources on the internet [7] and is used for web client request and response. The URL is made up of a sequence of strings, whereas some string has little semantic meaning. It is very tedious to derive semantic meaning from string in few URLs because joined words are incoherent.

A simple example for URL is given below:

https://www.google.co.in/search?q=wikipedia&sxsrf=ALiCzsax_704CVYK-3dXFzSX1dWRzU0B5dQ%3A1664036899393&source=hp&ei=IzAvY_i2Ffzh4-EPsaWWmAM

The above link has a scheme to predict the protocols used like HTTP or HTTPs. The second thing is the host name which will predict the machine that will have resources. It will have generic and country code top-level domain. In the above link, "co" is generic top-level domain, and "in" is the country code. Path refers to the basic information available in the host. In this example, path name is

search?q=wikipedia&sxsrf=ALiCzsax_704CVYK3dXFzSX1dWRzU0B-5dQ%3A1664036899393&source=hp&ei=IzAvY_i2Ffzh4-EPsaWWmAM

The next component is the query string where it is a part of the URL which allots value to the designated parameters.

Extracted features [8] will have varied formats and length. So, it will undergo pre-processing before getting into the later stages. Features extracted shall be in any form like text, binary format or may be in numeric value such as website sitemap depth or page content alikeness. The objective of pre-processing will reduce the complexity and training time of the network [22-26]. The extracted features which have text will require further steps in pre-processing.

  i. Cleaning: Eradicate the space, special symbols and unfamiliarized words.

ii. Words with vector representation: Here words in text are converted into vector format. Many tools are available for this conversion. It will tell us about the word similarity, so it decreases the language learning. It helps the network to discover the unseen words.
iii. While the data is unscaled or with huge value range or with large input it will decrease the framework convergence. Input standardization will speed up the training time and decrease the stuck occurrence in local optima. For input normalization, Gaussian distribution is utilized for zero mean and unit variance from zero to one.
iv. Webpage labelling with metadata tags will be helpful for the training process.
v. Datasets may have URLs redundancy and varied URLs with respect to the same host; while going for data deduplication process, we can eliminate the repeated URLs and URLs with same host.

8.3.2 Dataset Description

Some benchmarked datasets are used to evaluate the performance of the proposed system. The datasets used for the study includes normal dataset and phishing dataset. Two sets of datasets are used for the analysis of the proposed system. The first dataset is collected from Kaggle named as webpage phishing detection dataset and is available in the link: https://www.kaggle.com/datasets/shashwatwork/web-page-phishing-detection-dataset. The dataset has 11,400 URLs and 87 features. The features represent three different classes: 56 features denote URLs' structure and syntax, 24 features are taken from the content of each pages and 7 features are extracted by external services query. The dataset has equal distribution of both legitimate and phished URLs; each of them is 50% of the total size.

The next set of dataset used for the study is collected from the Phishtank dataset service provider. This provider has 5,000 phishing URLs that can be used for the model analysis and also provides dataset in CSV and JSON formats. The link to download dataset is https://www.phishtank.com/developer_info.php. Legitimate URLs' datasets are collected from New Brunswick University's dataset collection. Totally 5,000 legitimate URL links are taken for the analysis of the proposed system. The link for the dataset is available at https://www.unb.ca/cic/datasets/url-2016.html.

8.3.3 Performance Metrics

To predict the effectiveness of a proposed model, Accuracy, precision, recall, F-measure and error rate are taken into account for performance metrics evaluation.

Accuracy is defined as predicting the phishing and legitimacy rate of the total number of websites as shown in the below equation (8.2).

$$Accuracy = \frac{TP+TN}{TP+FP+TN+FN} \quad (8.2)$$

High recall is stated as a reduced number of phishing websites that are termed as legitimate in the below equation (8.3).

$$Recall = \frac{TP}{TP+FN} \quad (8.3)$$

Higher precision is defined as a reduced number of legitimate websites that are identified as phishing websites as shown in the below equation (8.4).

$$Precision = \frac{TP}{TP+FP} \quad (8.4)$$

F-measure termed as harmonic mean between precision and recall is shown in the below equation (8.5).

$$F-measure = 2 \cdot \frac{Precision \cdot Recall}{Precision + Recall} \quad (8.5)$$

Error Rate calculates the legitimacy rate or phishing from wrongly categorized websites as represented in (8.6).

$$Error\ rate = 1 - \frac{TP+TN}{P+N} \quad (8.6)$$

From the above equation FP, FN are the false positives, false negatives whereas TP and TN are true positives and true negatives of the model, respectively. High precision is recommended in the case when false positives are not favored, high recall is preferable in situations when false negatives are not considerable. Higher value of F-measure gives the better performance of our model.

8.4 System Evaluation

This section describes the performance of the proposed method (HTML and URL) in terms of various performance metrics. The different ML classifiers are implemented for evaluation of extracted features which are used in the proposed method.

Table 8.1 shows different extracted features for text like TF-IDF word level, TF-IDF N-gram level, TF-IDF character level, global to vector (GloVe) pre-trained word embedding, character sequences vectors, count vectors (bag-of-words), word sequences vectors, trained word embedding and implementation of various classifiers. The absolute aim of the designed system is to find out the better textual features suitable for the selected data. From the obtained results, it is observed that character level TF-IDF features provides best performance compared to other features with significant accuracy, precision, F-Score, Recall, and AUC using XG Boost and Deep Neural Network classifiers. Thus TF-IDF character level technique is implemented in this work to generate text features (F2) of the webpage.

Table 8.1 The different textual content features performance on D1 dataset with different classifiers.

Classifier	Textual data features	Precision (%)	F-score (%)	AUC (%)	Recall (%)	Accuracy (%)
XG Boost	TF-IDF word level	89.32	90.22	91.12	92.03	92.95
	TF-IDF N-gram level	88.65	89.53	90.43	91.33	92.25
	TF-IDF character level	89.90	90.80	91.71	92.62	93.55
	Word sequence vectors	83.49	84.32	85.16	86.02	86.88
	Count vectors	89.14	90.03	90.93	91.84	92.76
	Character sequence vectors	82.28	83.11	83.94	84.78	85.63

(Continued)

Table 8.1 The different textual content features performance on D1 dataset with different classifiers. (*Continued*)

Classifier	Textual data features	Precision (%)	F-score (%)	AUC (%)	Recall (%)	Accuracy (%)
NB	TF-IDF word level	85.35	86.20	87.06	87.93	88.81
	TF-IDF N-gram level	83.27	84.11	84.95	85.80	86.66
	TF-IDF character level	77.21	77.99	78.77	79.55	80.35
	Count vectors	83.45	84.28	85.12	85.97	86.83
	Word sequences vectors	63.52	64.15	64.80	65.44	66.10
	Trained word embedding	89.02	89.91	90.81	91.72	92.64
CNN	Character embedding	82.88	83.71	84.55	85.39	86.25
	Trained word embedding	90.47	91.37	92.28	93.21	94.14
RF	TF-IDF word level	86.80	87.67	88.54	89.43	90.32
	TF-IDF N-gram level	87.64	88.51	89.40	90.29	91.20
	TF-IDF character level	86.29	87.16	88.03	88.91	89.80
	Count vectors	86.67	87.53	88.41	89.29	90.19
	Word sequence vectors	82.38	83.20	84.03	84.87	85.72
	Character sequence vectors	80.31	81.11	81.92	82.74	83.57

(*Continued*)

Table 8.1 The different textual content features performance on D1 dataset with different classifiers. (*Continued*)

Classifier	Textual data features	Precision (%)	F-score (%)	AUC (%)	Recall (%)	Accuracy (%)
DNN	TF-IDF word level	87.95	88.83	89.72	90.62	91.52
	TF-IDF N-gram level	89.00	89.89	90.79	91.70	92.62
	TF-IDF character level	89.20	90.10	91.00	91.91	92.83
	Count vectors	88.36	89.25	90.14	91.04	91.95
	Word sequence vectors	54.80	55.35	55.90	56.46	57.03
	Character sequence vectors	77.17	77.95	78.73	79.51	80.31
LR	TF-IDF word level	86.54	87.40	88.28	89.16	90.05
	TF-IDF N-gram level	86.08	86.94	87.81	88.69	89.58
	TF-IDF character level	85.40	86.25	87.11	87.98	88.86
	Count vectors	87.71	88.59	89.47	90.37	91.27
	Word sequence vectors	56.43	56.99	57.56	58.14	58.72

Table 8.2 displays the experimental results with hyperlinks features. From the predicted results, it is evident that RF is a classifier better than other classifiers with an accuracy 83.09%, precision 78.37%, AUC 83.40%, recall 86.96% and F_Measure 82.45%, and it is evident that ensemble and XG Boost classifiers have obtained better accuracy of 83.00% and 81.29%, respectively.

In Table 8.3, we coordinated elements of HTML and URL (hyperlink and text) using different classifiers to check complementary behavior in phishing sites recognition. With the experimental outcomes, it is seen that LR classifier has adequate exactness and performance as far as the HTML highlights. Interestingly, NB classifier has great exactness, ACC, F1-Score, AUC, and review regarding consolidating every one of the highlights. RNN and gathering classifiers accomplished high precision, review, F1-Score, and AUC regarding URL-based highlights.

Table 8.2 Performance of the proposed AO-RNN hyperlink features on D1 with different classifiers.

Classifier	Precision (%)	AUC (%)	Recall (%)	F_Measure (%)	Accuracy (%)
RNN	78.37	83.40	86.96	82.45	83.09
NB	68.99	60.22	31.92	43.64	62.63
XG Boost	76.31	81.63	85.62	80.70	81.29
Ensemble	71.16	81.33	85.09	80.39	81.00
LR	69.74	67.99	56.23	62.27	68.99

Table 8.3 Various feature combinations on dataset D1.

Classifier	Features	Precision (%)	Recall (%)	F-score (%)	AUC (%)	Accuracy (%)
Ensemble	FURL	99.40	92.97	96.08	96.35	96.65
	FHTML	91.12	82.83	86.78	88.12	88.58
	FURL+HTML	94.83	88.73	91.68	92.44	92.75
RNN	FURL	99.53	93.06	96.18	96.44	96.74
	FHTML	91.68	82.80	87.02	88.35	88.83
	FURL + HTML	94.75	87.66	91.06	91.89	92.25
NB	FURL	82.22	22.31	35.11	59.51	62.68
	FHTML	66.33	88.45	75.81	75.23	74.11
	FHTM + URL	87.86	62.77	73.24	77.93	79.22
XG Boost	FURL	100.58	93.19	96.75	96.93	97.25
	FHTML	89.09	88.56	88.82	89.79	89.90
	FURL + HTML	99.26	95.51	97.34	97.55	97.73
LR	FURL	75.42	68.60	71.84	74.99	75.54
	FHTML	84.34	82.80	83.57	85.00	85.19
	FURL+HTML	78.49	69.43	73.68	76.82	77.45

In this examination, we contrast our methodology with existing anti-phishing approaches. Also the proposed methodology is assessed on benchmark dataset D26,13,30s in light of the four measurements utilized. The obtained examination results are displayed in Table 8.4. With the achieved outcomes, it is seen that the proposed methodology provides preferred execution over different methodologies examined in the survey, which displays the proficiency of distinguishing phishing sites over the current methodologies. Figure 8.1 shows the achievement of RNN with other approaches using dataset 1. Figure 8.6 shows the performance of RNN with other algorithms using dataset 1. Figure 8.7 shows the performance of RNN with various other procedures using dataset 2.

Table 8.4 Comparison of AO-RNN vs. Other Standard Approaches (for Dataset 1&2).

Dataset	Methods	Precision (%)	Recall (%)	F-score (%)	Accuracy (%)
Dataset 1	URLNET	96.86	90.51	93.59	94.38
	CNN	95.31	98.35	96.81	96.42
	AO-RNN	98.77	95.03	96.86	97.24
Dataset 2	URLNET	97.49	99.50	98.49	98.47
	CNN	98.53	98.91	98.72	98.74
	Deep CNN	87.14	88.44	87.74	90.55
	AO-RNN	98.50	99.54	99.01	98.97

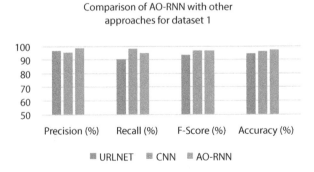

Figure 8.6 Performance of RNN with other approaches.

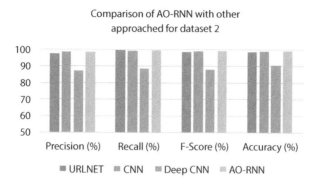

Figure 8.7 Performance of RNN with other approaches.

8.5 Conclusion

The phishing site appears to be like its harmless authority site, and the resistance is the means by which to recognize them. This work has introduced a clever anti-phishing method that includes various elements (URL, hyperlink, and text) that was not discussed in any of the other works. The approach introduced here is a totally client-side arrangement. These highlights proposed in this work are used on different AI calculations and it was found that RNN achieved the best execution. Our significant point is to plan a constant methodology, which produces lower misleading positive rate and higher evident negative rate. The outcomes display that our methodology accurately sifted the harmless website pages with a lower measure of harmless pages mistakenly delegated phishing. During the time spent phishing page arrangement, we build the dataset by removing the applicable and helpful elements from harmless and phishing pages.

References

1. Basit, A., Zafar, M., Liu, X., Javed, A. R., Jalil, Z., & Kifayat, K. (2021). A comprehensive survey of AI-enabled phishing attacks detection techniques. *Telecommunication Systems*, 76(1), 139-154.
2. Aleroud, Ahmed, and Lina Zhou. Phishing environments, techniques, and countermeasures: A survey. *Computers & Security* 68 (2017): 160-196.
3. Gillman, D., Lin, Y., Maggs, B., & Sitaraman, R. K. (2015). Protecting websites from attack with secure delivery networks. *Computer*, 48(4), 26-34.

4. Basit, A., Zafar, M., Liu, X., Javed, A. R., Jalil, Z., & Kifayat, K. (2021). A comprehensive survey of AI-enabled phishing attacks detection techniques. *Telecommunication Systems*, 76(1), 139-154.
5. Feng, T., & Yue, C. (2020, June). Visualizing and interpreting RNN models in URL-based phishing detection. In *Proceedings of the 25th ACM Symposium on Access Control Models and Technologies* (pp. 13-24).
6. Lin, W. H., Lin, H. C., Wang, P., Wu, B. H., & Tsai, J. Y. (2018, April). Using convolutional neural networks to network intrusion detection for cyber threats. In *2018 IEEE International Conference on Applied System Invention (ICASI)* (pp. 1107-1110). IEEE.
7. Elnagar, S., & Thomas, M. (2018, July). A cognitive framework for detecting phishing websites. In *Proceedings of the International Conference on Advances on Applied Cognitive Computing, Las Vegas, NV, USA* (pp. 60-61).
8. Somesha, M., Pais, A. R., Rao, R. S., & Rathour, V. S. (2020). Efficient deep learning techniques for the detection of phishing websites. *Sādhanā*, 45(1), 1-18.
9. Anand Joseph Daniel, G. Reshma, C. Selvarani (2021), To Detecting Phishing Attacks Using Natural Language Processing and Machine Learning, *International Journal of Advanced Research in Computer and Communication Engineering* Vol. 10, Issue 7, July 2021.
10. Mahmoud Khonji, Youssef Iraqi, and Andrew Jones. (2013) Phishing Detection: A Literature Survey. IEEE.
11. B. Wardman, T. Stallings, G. Warner and A. Skjellum, High-performance content-based phishing attack detection, *2011 eCrime Researchers Summit*, 2011, pp. 1-9, doi: 10.1109/eCrime.2011.6151977.
12. M. N. Alam, D. Sarma, F. F. Lima, I. Saha, R.-E.- Ulfath and S. Hossain, "Phishing Attacks Detection using Machine Learning Approach," *2020 Third International Conference on Smart Systems and Inventive Technology (ICSSIT)*, 2020, pp. 1173-1179, doi: 10.1109/ICSSIT48917.2020.9214225.
13. Chen, W., Zhang, W., Su, Y. (2018). Phishing Detection Research Based on LSTM Recurrent Neural Network. In: Zhou, Q., Gan, Y., Jing, W., Song, X., Wang, Y., Lu, Z. (eds.) *Data Science. ICPCSEE 2018. Communications in Computer and Information Science*, vol 901. Springer, Singapore.
14. Sountharrajan, S., Nivashini, M., Shandilya, S.K., Suganya, E., Bazila Banu, A., Karthiga, M. (2020). Dynamic Recognition of Phishing URLs Using Deep Learning Techniques. In: Shandilya, S., Wagner, N., Nagar, A. (eds.) *Advances in Cyber Security Analytics and Decision Systems. EAI/Springer Innovations in Communication and Computing*. Springer, Cham.
15. F. Salahdine, Z. El Mrabet and N. Kaabouch, Phishing Attacks Detection: A Machine Learning-Based Approach, *2021 IEEE 12th Annual Ubiquitous Computing, Electronics & Mobile Communication Conference (UEMCON)*, 2021, pp. 0250-0255, doi: 10.1109/UEMCON53757.2021.9666627.
16. I. Saha, D. Sarma, R. J. Chakma, M. N. Alam, A. Sultana and S. Hossain, Phishing Attacks Detection using Deep Learning Approach. In *2020*

Third International Conference on Smart Systems and Inventive Technology (ICSSIT), 2020, pp. 1180-1185.
17. Basnet, R., Mukkamala, S., Sung, A.H. (2008). Detection of Phishing Attacks: A Machine Learning Approach. In Prasad, B. (ed.) *Soft Computing Applications in Industry. Studies in Fuzziness and Soft Computing*, vol 226. Springer, Berlin, Heidelberg.
18. Alfredo Cuzzocrea, Fabio Martinelli, Francesco Mercaldo, Applying Machine Learning Techniques to Detect and Analyze Web Phishing Attacks, *Proceedings of the 20th International Conference on Information Integration and Web-based Applications & Services*, November 2018, pp. 355–359. https://doi.org/10.1145/3282373.3282422.
19. Dutta AK. Detecting phishing websites using machine learning technique. *PLoS One*. 2021 Oct 11;16(10):e0258361. doi: 10.1371/journal.pone.0258361.
20. Jain A.K., Gupta B.B. "PHISH-SAFE: URL Features-Based Phishing Detection System Using Machine Learning", *Cyber Security. Advances in Intelligent Systems and Computing*, vol. 729, 2018, https://doi.org/10.1007/978-981-10-8536-9_44
21. Zhao, J., Wang, N., Ma, Q., Cheng, Z.: Classifying malicious URLs using gated recurrent neural networks. In: *Proceedings of the 12th International Conference on Innovative Mobile and Internet Services in Ubiquitous Computing*. pp. 385-394. IMIS'18, Springer, Cham (July 2018).
22. Sahingoz, O. K., Buber, E., Demir, O. & Diri, B. Machine learning based phishing detection from URLs. *Expert Syst. Appl.* 2019 (117), 345-357 (2019).
23. Rao, R. S., Vaishnavi, T. & Pais, A. R. Catch Phish: Detection of phishing websites by inspecting URLs. *J. Ambient. Intell. Humanized Computing*. 11, 813-825 (2019).
24. Chatterjee, M., & Namin, A.S. Detecting phishing websites through deep reinforcement learning. in *2019 IEEE 43rd Annual Computer Software and Applications Conference (COMPSAC)*. 978-1-7281-2607-4/19/$31.00 ©2019 IEEE. (IEE Computer Society, 2019).
25. S., Priyanka, *et al*. Hindrance Detection and Avoidance in Driverless Cars Through Deep Learning Techniques. In *Advances in Deep Learning Applications for Smart Cities*, edited by Rajeev Kumar and Rakesh Kumar Dwivedi, IGI Global, 2022, pp. 69-100. https://doi.org/10.4018/978-1-7998-9710-1.ch005
26. Aljofey, A., Jiang, Q., Qu, Q., Huang, M. & Niyigena, J.-P. An effective detection approach for phishing websites using URL and HTML features, Nature. com (2022).

9

Security Risks and Its Preservation Mechanism Using Dynamic Trusted Scheme

Geetanjali Rathee[1*], Akshay Kumar[2], S. Karthikeyan[3] and N. Yuvaraj[4]

[1]Department of CSE, Netaji Subhas University of Technology, Dwarka, New Delhi, India
[2]Department of CSE, Jaypee Institute of Information Technology, Noida, U.P., India
[3]Department of CSE (AI&ML), KPR Institute of Engineering and Technology, Coimbatore, India
[4]Department of AI&DS, KPR Institute of Engineering and Technology, Coimbatore, India

Abstract

The dynamic decision making and accurate decisions while communicating information in real-time applications lead to various security threats in the network. A number of intruders may steal the ideal nodes and manipulate them to their benefit, which may cause further drastic degradation of any organizational growth in the market. The objective of this paper is to propose a secure and trusted communication mechanism by highlighting their various risks and preservation methods for wireless communication. The proposed method highlights the importance of trust-based computation while ensuring an efficient security and preservation in the network. The method is further verified over various security metrics against traditional scheme.

Keywords: Dynamic trusted approach, security risks, privacy scheme, secure IoT, trust-based devices

---

Corresponding author: geetanjali.rathee123@gmail.com

S. Sountharrajan, R. Maheswar, Geetanjali Rathee, and M. Akila (eds.) Wireless Communication for Cybersecurity, (199–216) © 2023 Scrivener Publishing LLC

9.1 Introduction

Nowadays, the generation of a huge amount of information can be easily handled using various modern techniques or technologies in the environment. The millions and trillions of information can be easily handled, analyzed, processed and stored through online databases and systems such as cloud systems, Hadoop, servers, etc. [1, 2]. The intervention of human effort is further reduced by replacing the human loads with smart/intelligent automation systems. Along with several advantages of modern techniques, these technologies also bring a huge number of other types of risks and challenges in the network. Security is considered as one of the types of crucial issue that is very difficult to manage while providing the communication through various smart devices in the network [3, 4]. It is crucial to ensure the security of intelligent devices while sharing the information among each other in the network. The dynamic systems are the ones where devices are heterogenous in nature and try to communicate among each other and ask for message transmission in the network. The number of security risks arise during the dynamic nature of devices such as denial of service, man-in-middle attack, distributed network, etc. [5, 6]. in addition, it is necessary to determine various preserving mechanisms to ensure a reliable communication among devices in the network. There are a number of security schemes to ensure security, such as encryption techniques, ticket-based techniques, trust-based techniques, etc. In this paper, we will discuss a number of trust-based security mechanisms.

9.1.1 Need of Trust

The involvement of smart devices automation while making any real-time decision lends more accuracy to the environment. However, the involvement of any type of threat of an intruder that is not allowed to proceed in its communication may further drastically affect the entire network. A number of security risk handlers, mechanisms, algorithms and mechanisms have been proposed by several researchers/scientists; however, other security risks automatically degrade the entire system performance [7, 8]. For instance, the alteration of ideal nodes in the network by the intruders may further increase the traffic congestion, overhead, storage and other types of risks in the network. Therefore, it is necessary to further improve the entire set of the communication process handled through intelligent systems or IoT devices.

RISK AND PRESENTATION USING DYNAMIC TRUST 201

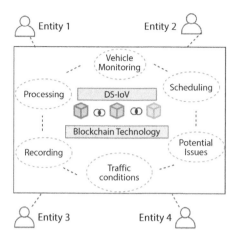

Figure 9.1 A secured smart society framework having intelligent devices in the network [7].

Figure 9.1 represents the overall communication process in a smart automation system where a number of IoT devices are involved in the network while performing an efficient communication network in the system. Among various types of security models, trust-based security is considered as one of the efficient communication ways in the network. The trust-based computation improves the security process without affecting the cost, storage and much communication system in the network [9, 10].

9.1.2 Need of Trust-Based Mechanism in IoT Devices

Though a number of security schemes have been proposed by various researchers/scientists, it is still necessary to propose a reliable, legitimate and transparent communication mechanism in the system. Figure 9.2 represents the trust-based mechanism in intelligent devices where a number of ambient systems during the communication in various heterogenous systems may lose security. The integration of indirect or any other trust-based computation either with reinforcement learning or any other device may further improve the security in the network.

9.1.3 Contribution

The main aim of this paper is to propose a secure and efficient communication procedure using dynamic trust-based computation that is used to detect/identify the trust value of each communicating device [11, 12]. The

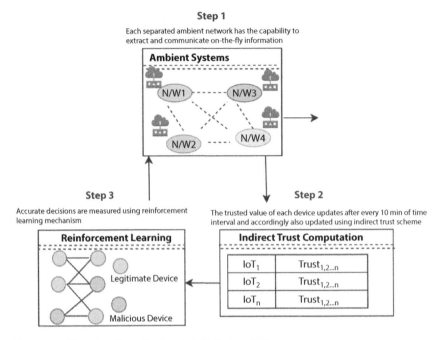

Figure 9.2 Trust-based mechanism in IoT devices [27].

proposed mechanism computes the updated trust degree of the entire communication network using various security metrics. The remaining organization of the paper is as follows. Section 9.2 describes the related work of various security models/methods proposed by several scientists. Section 9.3 determines a proposed communication model using trust-based computation that further updates the trust rate of each communicating device. Further, section 9.4 determines the performance analysis of the proposed framework for validating the scenario. Finally, section 9.5 concludes the entire work along with its future research directions.

9.2 Related Work

This section describes a number of security approaches and methods proposed by various academicians, scientists and authors for ensuring a secure and effective communication process in the network [13, 14]. Table 9.1 illustrates the number of security methods/algorithms and schemes along with their limitations.

Table 9.1 Literature survey.

Author	Technique	Measuring parameter	Limitation
Zeke et al. [15]	Secure communication process using fuzzy systems	Determined the secure evaluation factor suing weight coefficient by computing the matrix and various security levels	Suffer from communication delay
Zhong et al. [16]	Index confirmation process	Detailed the concept by deducing the mathematical curves and formulas on various classification regions	Energy consumption while tracking the mechanism again and again
Angelogianni et al. [17]	Systematic identification mechanism	Evaluated the scenarios into three distinct categories	Authentication process leads to delay
Guo et al. [18]	Fuzzy and AHP based evaluation method for assessment	Considered the advantages of both the existing frameworks such as AHP and fuzzy for ensuring accurate results and operability during the risk assessment.	Issues during dynamic behavior of the network
Liang et al. [19]	Risk evaluation for industrial sensor networks	Studied the issues of security and risks for wireless sensors in industries	Need to consider accuracy
Wu et al. [20]	Cross-project security	Proposed a hybrid security based on uncertainty and text similarity in the network	Needed to analyze the computation delay of trust

Zeke *et al.* [15] have proposed a monitoring system for ensuring a secure communication process using fuzzy systems in the network. The authors have determined the secure evaluation factor using weight coefficient by computing the matrix and various security levels. The proposed mechanism provided various methods and ideas for ensuring a secure communication process for wireless energy stations. Zhong *et al.* [16] have proposed an index confirmation process for identifying the eavesdropping in a wireless communication process. The authors have detailed the concept by deducing the mathematical curves and formulas on various classification regions. The proposed framework is claimed to reduce the eavesdropping probability over existing method. Angelogianni *et al.* [17] have proposed a systematic identification mechanism for detecting the vulnerabilities while considering present legislative frameworks. The proposed risk management and assessment system evaluated the scenarios into three distinct categories. The evaluated results validated the proposed framework against state-of-art cellular schemes. Guo *et al.* [18] have proposed a fuzzy and AHP-based evaluation method for assessment of the risk in the system. The authors have considered the advantages of both the existing frameworks such as AHP and fuzzy for ensuring accurate results and operability during the risk assessment.

In addition, Shabisha *et al.* [21] have projected an enhanced and new security mechanism for determining the emergency issues in healthcare systems. The authors have proposed a key agreement and an authenticated security mechanism by relying on the symmetric key schemes for measuring the security and anonymity on the nodes. The authors have further developed a commercial off the shelf system while transmitting the information by generating the warning and emergency alarms. Further, Zhang *et al.* [22] have studied numerous controlling strategies including coefficient and association designs from the network perspectives. The authors have projected the online and offline controlling strategies for accessing the channel information. They have projected a dual composition transforming function for designing the distributed controlling scheme.

Though a number of schemes have been proposed, it is further necessary to focus on the accuracy and computational steps of trust along with reduced transparency time using blockchain system. This paper proposed a secure and trusted communication mechanism using blockchain-based adaptive and comprehensive trust computation of each DS that is further verified against accuracy, computational delays and probability attacks of each communicating devices [23-25].

9.3 Proposed Framework

9.3.1 Dynamic Trust Updation Model

In this paper, we are proposing a dynamic trust updations mechanism that is based upon Bayesian formulation using beta distribution method as proposed in [20]. In addition, the trusted mechanism is further integrated with blockchain technology in order to strengthen the risk analysis and detection by providing transparency during distribution or transmission of messages in the network. Now, the trust of a node is computed by maintaining a data structure having a number of variables such as α, β, p, q. Each node i will maintain a trust degree of any node j in its trust table T as T_{ij} that represents the trust degree of node j maintained by node j in its trust table. In addition, the trust degree keeps on updating depending upon its internal communication activities and behavior that can be examined and represented as Internal Behavior (IB_{ij}) trust degree. The interaction (I) among internal behavior of each node and its table degree is represented as:

$$T_{ij} = I(IB_{ij}, T_{ij})$$

Where, IB_{ij} is further mapped to a pair of (p, q) that represents the rating allocated to the node j by node i depending upon its recent activity. In addition, I(.) is the updating trust degree due to interaction among nodes and communication among each other. I(.) is responsible to update the trust degree of node j depending upon its recent communication behavior and activeness in the network.

Further, the proposed model uses the Bayesian formulation along with beta distribution method inspired by the work of Ganeriwal *et al.* The Bayesian formulation using beta distribution scheme separates the trustworthy and non-trustworthy nodes in α and β where trust table can be further represented as:

$$T_{ij} = \beta(\alpha_i + 1, \beta_j + 1)$$

Where, α_i, β_j determines the trustworthy and non-trustworthy communication interactions among node I and node j.

Furthermore, the trust degree keeps on updating using Internal Behavior (IB_{ij}) where p and q are considered as some if the integers that increase or decrease the rating of each communicating node in the network.

The updated trust degree corresponding to node j by node I can be further represented as:

$$T_{ij} = \beta(\alpha_i + 1 + p, \beta_j + 1 + q) \qquad (9.1)$$

Where, the two parameters α_i and β_j values can be further defined as:

$$\begin{cases} \alpha_j^u = (e * \alpha_i) + p \\ \beta_j^u = (e * \beta_j) + q \end{cases} \qquad (9.2)$$

Where, e termed as changing weights denoted as e between (0,1).

Now, in order to keep a surveillance of changed trust degrees by the communicating nodes can be further traced using blockchain mechanism. The section below determines the blockchain benefits while maintaining or changing the trust degrees of each communicating node.

Here, we are proposing a dynamic trust updations mechanism that is based on Bayesian formulation using beta distribution method as proposed in [19]. In addition, the trusted mechanism is further integrated with blockchain technology in order to strengthen the risk analysis and detection by providing the transparency during distribution or transmission of messages in the network. Now, the trust of a node is computed by maintaining a data structure having a number of variables such as α,β,p,q. Each node i will maintain a trust degree of any node j in its trust table T as T_ij that represents the trust degree of node j maintained by node j in its trust table. In addition, the trust degree keeps on updating depending upon its internal communication activities and behavior that can be examined and represented as Internal Behavior (IB_{ij}) trust degree.

The other classifiers such as KNN, Naïve Bayes, etc., can be used to determine the legitimacy of each communicating device instead of Bayesian formulation method; however, the mentioned performance metrics are much accurately determined because of updations of trust values continuously. Furthermore, the trust degree keeps on updating using Internal Behavior (IB_{ij}) where p and q are considered as some of the integers that increase or decrease the rating of each communicating node in the network.

The out-performance of proposed framework is due to its changing trust degree upon each communication process that updates the internal behavior of each node in the network. The altered node can be easily identified by the proposed approach because of its changed trust degree.

9.3.2 Blockchain Network

Algorithm 1

Begin
Step 1: *Input*: A set of communicating nodes in a network as N_n
Output: Device is trustworthy or non-trustworthy.
Given: Various categories of nodes along with a blockchain network.
Step 2: a) establish a network N_n where node j trust degree is maintained by node I and so on
    For all nodes N_n = {where N=1,2,....n}
        For i=1 to n then
            Compute the trust degree of each node using Bayesian formulation and beta distribution scheme
            If (Device is trustworthy) then
                Maintain a blockchain and keep surveillance of its behavior
            Else
                Node will not be allowed for further communication in the network
            End if
    b) Maintain a Blockchain of trustworthy nodes in the network.
Step 3: Each trustworthy node is surveillance with their trust degrees using blockchain network
    End For
    End For

The Bayesian formulation using beta distribution scheme separates the trustworthy and non-trustworthy nodes in α and β where trust table can be further represented. Each node i will maintain a trust degree of any node j in its trust table T as T_{ij} that represents the trust degree of node j maintained by node j in its trust table. In addition, the trust degree keeps on updating depending upon its internal communication activities and behavior that can be examined and represented as Internal Behavior (IB_{ij}) trust degree. The blockchain network maintains a block of chain having all the communicating nodes in the network. The nodes that are communicating and their weights are changing depending upon their internal behavior can be easily traced using blockchain architecture. The blockchain network maintains a block of nodes having their trust degrees and neighboring nodes information in the network. The process of integration of dynamic updations of trust degrees with blockchain network can be illustrated in Algorithm 1.

The presented algorithm 1 represents the device's legitimacy by computing the trust values during data transmission in the network. The

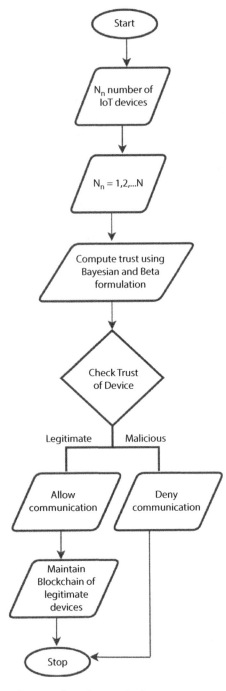

Figure 9.3 Flowchart of proposed mechanism [26].

trust values are computed using Bayesian formulation for computing the behavior of each device. Further, in order to maintain the continuous surveillance of the communicated nodes, the devices are maintained in a blockchain network. The word dynamic with the updations is used in case of changing the trust degrees upon each communication between node i and node j. The trust values are changing based upon their internal communicational behavior and activeness ratio in the network. The main aim of this manuscript is to ensure the security amongst users that utilizes the network services or provide security of stored information. Further, the flowchart of the proposed mechanism is presented in Figure 9.3. The presented flowchart determines the explanation of the proposed mechanism where devices are communicated while computing their trust values. The devices that have significant trust values are defined as legitimate and allowed to perform communication and transmission of information in the network. In order to reduce storage overhead and to improve efficiency of the network, we have maintained a blockchain for different purposes. The size of data in communicated device is considered as 128 bits that is further checked by determining their trust values using Bayesian formulation scheme. The network is updated after a specific amount of interval for their continuous assessment that is further maintained through blockchain. If the trust degree information sent by other nodes are corrupted, in that case for the being time the legitimate device can be further identified as altered. However, the previous history interaction and their continuous behavior in the network while forwarding the information may further update the trust degree of that device. In addition, all the devices along with their trust degrees are added in the blockchain that keeps the entire history of their trust values while communicating the information in the network.

9.4 Performance Analysis

The performance evaluation of the proposed mechanism that is simply defined as the integration of trust degrees and blockchain network is implemented and verified using MATLAB simulation. The network size of 50 nodes is considered for validating the proposed scenario against a traditional approach over various security metrics.

9.4.1 Dataset Description and Simulation Settings

Having a synthesized dataset where some of the nodes are selected as trustworthy and some of them are acted as non-trustworthy by externally is used to measure the outperformance of the network. The simulation is run in a network size of 700 * 700 m of area size, 50 number of communicating nodes starting from 5-50 and a running time of 60s with a channel capacity of 2.5 Mbits. In addition, IEEE 802.11 MAC protocol is used to maintain the communication with some defined initial values such as α_i, β_j as 0,0 and value of e as 0.95 during the network establishment. Initially the trust value is randomly distributed among (0,1) that can be further updated by computing the trust degree of each communicating node in the network.

9.4.2 Traditional Method and Evaluation Metrics

The proposed framework is analyzed against Zhong *et al.* [8] (known as baseline approach) in which the authors have proposed an index confirmation process for identifying the eavesdropping in a wireless communication process. The authors have detailed the concept by deducing the mathematical curves and formulas on various classification regions. The proposed framework is claimed to reduce the eavesdropping probability over existing method.

In addition, the evaluation metrics used to verify the outperformance of proposed framework over existing work is defined as node alteration rate, accuracy, and delay. The in-depth definition of each metrics is determined as below:

Alteration rate: The alteration rate is termed as the number of nodes that can be easily altered from trustworthy to non-trustworthy by the intruders during the communication process in the network.
Accuracy: The accuracy is generally used to measure the complexity of the proposed mechanism regarding how much time and effort the approaches need to accurately determine the trustworthiness or internal behavior of each communicating node.
Delay: It is defined as the amount of time required to determine the ideal nature or activity of each communicating node in the network.

9.5 Results Discussion

Figure 9.4 presents the graph of alteration rate that clearly determines the outperformance of the network over the existing approach. The outperformance of the proposed framework is due to its changing trust degree upon each communication process that updates the internal behavior of each node in the network. The altered node can be easily identified by

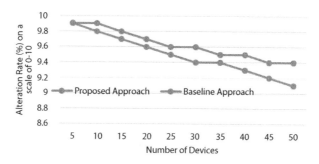

Figure 9.4 Alteration rate.

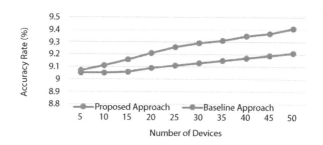

Figure 9.5 Accuracy rate.

the proposed approach because of its changed trust degree. Figure 9.5 depicts the accuracy rate of the proposed framework against the existing mechanism.

The accuracy of trust degree is due to its blockchain integration where the nodes are integrated depending upon its activeness in the environment. The nodes having higher trust degree provide more accurate results that will always be in surveillance using blockchain network.

The nodes having lesser trust degrees will never take part in communication process and will never be part of blockchain network. Finally, Figure 9.6 determines the delay, which means the amount of time required to determine the accuracy and trustworthiness of the node in the network. The nodes having higher trust degrees are also the part of blockchain network and keep on surveillance by the environment. The delay in identifying the trustworthy nodes in the proposed mechanism is much less as compared to the existing approach because of involvement of blockchain network.

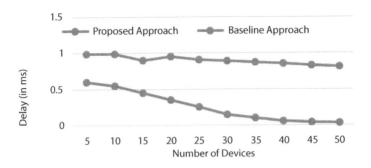

Figure 9.6 Delay.

9.6 Empirical Analysis

The performance analysis can be further measured using an empirical analysis method where the performing metrics are analyzed based upon their complexity and communication delay. The text below determines the number of empirical analysis schemes for further analyzing the proposed and existing methods.

- Energy consumption: The transmission of information and communication among the devices leads to consumption of energy. The devices consume an amount of energy while communicating and emitting the energy loss in the environment.
- Communication overhead: The communication overhead is considered as another significant metric while analyzing the performance of a network. The large amount of security schemes or methods may result in huge computations that may further delay the response by the network.
- Response delay: the response delay metric determines how much time a systems needs to respond to the requested input from the user. The device having huge communication and communication overhead may lead to a large delay and late response in the network.
- Falsification: the information falsification where intruders successfully hack the legitimate devices ID and try to authenticate themselves in the network with the means of their own benefits.

All these metrics can be further analyzed for both proposed and existing approaches to measure the overall outperformance of the proposed phenomenon. By considering all the metrics, it is further analyzed that the proposed mechanism will always perform better as compared to the existing approach because the proposed approach does not have huge communication and computation loads to respective devices. The legitimacy of each device is analyzed using one single approach and reduces the energy consumption and overhead at the devices. In addition, the blockchain maintenance may further reduce the response delay and falsification threat because of continuous surveillance in the network.

9.7 Conclusion

A dynamic and updated trust-based communication and secure transmission process among wireless nodes process is presented in the paper. The trust degree of each communicating node is determined using Bayesian function having beta distribution. In addition, the communicating nodes are further surveilled by the blockchain network to increase its transparency in the network. The proposed mechanism is efficiently validated and verified against the existing approach over various security metrics such as alteration rate, accuracy and delay via communicating the information among various nodes in the network. The proposed framework out-performance can be clearly seen through various graphs and computed results using MATLAB simulation. The proposed framework can be further extended by including other security threats related to wireless communication such as Sybil attack, denial of service, throughput and end-to-end delay in the future communication.

References

1. Joshi, C., & Singh, U. K. (2017). Information security risks management framework–A step towards mitigating security risks in university network. *Journal of Information Security and Applications, 35,* 128-137.
2. Dewett, T., & Jones, G. R. (2001). The role of information technology in the organization: a review, model, and assessment. *Journal of Management, 27*(3), 313-346
3. Rathee, G., Garg, S., Kaddoum, G., Choi, B. J., Hassan, M., & Alqahtani, S. A. (2022). TrustSys: Trusted Decision Making Scheme for Collaborative Artificial Intelligence of Things. *IEEE Transactions on Industrial Informatics.*

4. Zaalouk, A., Khondoker, R., Marx, R., & Bayarou, K. (2014, May). OrchSec: An orchestrator-based architecture for enhancing network-security using network monitoring and SDN control functions. In *2014 IEEE Network Operations and Management Symposium (NOMS)* (pp. 1-9). IEEE.
5. Srinivasan, K., Rathee, G., Raja, M. R., Jaglan, N., Mahendiran, T. V., & Palaniswamy, T. (2022). Secure multimedia data processing scheme in medical applications. *Multimedia Tools and Applications*, 81(7), 9079-9090.
6. Sandberg, H., Amin, S., & Johansson, K. H. (2015). Cyberphysical security in networked control systems: An introduction to the issue. *IEEE Control Systems Magazine*, 35(1), 20-23.
7. Rathee, G., Saini, H., Kerrache, C. A., & Herrera-Tapia, J. (2022). A Computational Framework for Cyber Threats in Medical IoT Systems. *Electronics*, 11(11), 1705.
8. Adnane, A., Bidan, C., & de Sousa Júnior, R. T. (2013). Trust-based security for the OLSR routing protocol. *Computer Communications*, 36(10-11), 1159-1171.
9. Fang, W., Cui, N., Chen, W., Zhang, W., & Chen, Y. (2020). A trust-based security system for data collection in smart city. *IEEE Transactions on Industrial Informatics*, 17(6), 4131-4140.
10. Rathore, M. S., Poongodi, M., Saurabh, P., Lilhore, U. K., Bourouis, S., Alhakami, W., ... & Hamdi, M. (2022). A novel trust-based security and privacy model for Internet of Vehicles using encryption and steganography. *Computers and Electrical Engineering*, 102, 108205.
11. Gupta, S., Dhurandher, S. K., Woungang, I., Kumar, A., & Obaidat, M. S. (2013, October). Trust-based security protocol against blackhole attacks in opportunistic networks. In *2013 IEEE 9th International Conference on Wireless and Mobile Computing, Networking and Communications (WiMob)* (pp. 724-729). IEEE.
12. Coates, G. M., Hopkinson, K. M., Graham, S. R., & Kurkowski, S. H. (2008). Collaborative, trust-based security mechanisms for a regional utility intranet. *IEEE Transactions on Power Systems*, 23(3), 831-844
13. Rathee, G., Ahmad, F., Iqbal, R., & Mukherjee, M. (2020). Cognitive automation for smart decision-making in industrial internet of things. *IEEE Transactions on Industrial Informatics*, 17(3), 2152-2159.
14. Lin, H., Xu, L., & Gao, J. (2009, April). A New Security Mechanism for SIP-Based VoIP over WMNs. In *2009 International Conference on Networks Security, Wireless Communications and Trusted Computing* (Vol. 2, pp. 330-333). IEEE.
15. Zeke, L. I., Zewen, C. H. E. N., Chunyan, W. A. N. G., Zhiguang, X. U., & Ye, L. I. A. N. G. (2020, August). Research on security evaluation technology of wireless access of electric power monitoring system based on fuzzy. In *2020 IEEE 3rd International Conference on Computer and Communication Engineering Technology (CCET)* (pp. 318-321). IEEE.

16. Zhong, X., Fan, C., & Zhou, S. (2022). Eavesdropping area for evaluating the security of wireless communications. *China Communications, 19*(3), 145-157.
17. Angelogianni, A., Politis, I., Mohammadi, F., & Xenakis, C. (2020). On identifying threats and quantifying cybersecurity risks of mnos deploying heterogeneous rats. *IEEE Access, 8*, 224677-224701.
18. Guo, C., Wang, X., & Chu, P. (2021, March). Fuzzy AHP-Based Security Evaluation for Wireless Integrated Access System. In *2021 International Conference on Intelligent Transportation, Big Data & Smart City (ICITBS)* (pp. 554-557). IEEE.
19. Liang, L., Liu, Y., Yao, Y., Yang, T., Hu, Y., & Ling, C. (2017, April). Security challenges and risk evaluation framework for industrial wireless sensor networks. In *2017 4th International Conference on Control, Decision and Information Technologies (CoDIT)* (pp. 0904-0907). IEEE.
20. Wu, X., Fu, W., Mu, D., Mao, D., Zhang, H., & Zheng, W. (2020, August). Improving the Security of Wireless Network Through Cross-project Security Issue Prediction. In *2020 IEEE/CIC International Conference on Communications in China (ICCC)* (pp. 1179-1184). IEEE.
21. P. Shabisha, C. Sandeepa, C. Moremada, N. Dissanayaka, T. Gamage, A. Braeken, K. Steenhaut, M. Liyanage, Security enhanced emergency situation detection system for ambient assisted living, *IEEE Open Journal of the Computer Society* 2 (2021) 241-259.
22. L. Zhang, G. Feng, S. Qin, Y. Sun, B. Cao, Access control for ambient backscatter enhanced wireless internet of things, *IEEE Transactions on Wireless Communications*.
23. Malik, V., & Singh, S. (2019). Security risk management in IoT environment. *Journal of Discrete Mathematical Sciences and Cryptography, 22*(4), 697-709.
24. Agarwal, S., Oser, P., & Lueders, S. (2019). Detecting IoT devices and how they put large heterogeneous networks at security risk. *Sensors, 19*(19), 4107.
25. Alladi, T., Chamola, V., Sikdar, B., & Choo, K. K. R. (2020). Consumer IoT: Security vulnerability case studies and solutions. *IEEE Consumer Electronics Magazine, 9*(2), 17-25.
26. Rathee, G., Garg, S., Kaddoum, G., & Choi, B. J. (2020). Decision-making model for securing IoT devices in smart industries. *IEEE Transactions on Industrial Informatics, 17*(6), 4270-4278.
27. Rathee, G., Kerrache, C. A., & Calafate, C. T. (2022). An Ambient Intelligence approach to provide secure and trusted Pub/Sub messaging systems in IoT environments. *Computer Networks, 218*, 109401.

10

6G Systems in Secure Data Transmission

A.V.R. Mayuri*, Jyoti Chauhan, Abhinav Gadgil, Om Rajani and Soumya Rajadhyaksha

School of Computing Science and Engineering, VIT Bhopal University, Bhopal-Indore Highway, Kotrikalan, Sehore, Madhya Pradesh, India

Abstract

With the ever-growing increase in the technologies, the newer generation always has their eyes on the next level of every pre-existing technology that was ever made. Along with the improvement of 5G technologies, the industries and academia have commenced having their eyes on the 6th generation wireless network technology (6G), which is anticipated to have higher and numerous threat coping mechanisms. As the deployment of 5G networks have grown over the years, the number of limitations has also significantly increased. 6G networks are expected to have extended connectivity which the current generation might have not even thought of. The upcoming and cutting-edge technology together with post-quantum cryptography, artificial intelligence (AI), machine learning (ML), more advantageous edge computing, molecular communication, THz, visible light communication (VLC), and distributed ledger (DL) technology together with blockchain, is said to shape the spine of 6G networks. One critical element within the fulfillment of 6G can be security.

New novel authentication, encryption, get admission to control, communication, and malicious activity detection, new safety techniques are important to ensure trustworthiness and privacy of the future networks. On the basis of 5G technique, 6G will have a profound impact on ubiquitous connectivity, holographic connectivity, deep connectivity and practical connectivity. As 6G is expected to become an additional open network to 5G, the inside and outside of the network may become increasingly blurred. Therefore, current network security methods, equivalent to IPsec, firewall, intrusion detection system (IDSs) etc. that enforce security for network edge purposes will not be robust enough. To mitigate this limitation, 6G security design must support the fundamental security principle of Zero Trust (ZT) within the mobile communication network.

---

*Corresponding author: mayuri.avr@vitbhopal.ac.in

S. Sountharrajan, R. Maheswar, Geetanjali Rathee, and M. Akila (eds.) *Wireless Communication for Cybersecurity*, (217–238) © 2023 Scrivener Publishing LLC

Keywords: 6G systems, 6G security, 6th generation, wireless technology, holographic connectivity, secure data transmission

10.1 Introduction

6G (sixth generation) serves as the latest technology that uses wireless networks for cellular networks with higher frequencies and large coverage area. It's not implemented on a user level yet, but research has already been done for a more ubiquitous and reliable internet presence across all cellular networks. The 6G networks will be able to operate on higher frequencies than 5G networks. This will mean that the way transceiver boxes are manufactured will also be changed, which will result in new architecture for 6G networks. As the rule states, transmission rate requires a tighter distribution of cells over a network. It can be anticipated that 6G can be used to harness the power of 5G and improve it tenfold to deliver ultra-fast speeds, exceeding device capacity and next-to-no latency. The fundamental ingredient of the 6G network is that it's expected to selectively use different frequencies to measure absorption and adjust frequencies accordingly. This is because the atoms and molecules in a substance emit distinctive frequencies but their emission and absorption frequencies are the same for any substance. 6G will have big implications for many government and industry approaches to public safety and critical asset protection such as terahertz (THz) band, AI, optical wireless communication (OWC), 3D networking, unmanned aerial vehicles (UAV), and wireless power transfer. 6G networks are bound to have superior performance since they have to be evaluated on a lot more metrics and their measures are on a higher scale than 5G networks. In India, N77 and N78 are the most popular 5G bands that are in smartphones. But a higher 5G band ranges between 24-47GHz. This provides a maximum data throughput of 18-20 Gbps for 5G [6]. Because 6G networks run at many frequencies to adjust with the use of THz and optical frequency bands, this speed might be exponentially improved. With these high frequency bands, the data rate can reach Gbps at the user level. As a result, the area traffic throughput can exceed 1 Gbps/m2. To compensate for the 100-fold increase in data rate, spectrum efficiency can grow by 3–5 times, while energy efficiency must increase by more than 100 times. The use of artificial intelligence (AI) can help with the administration of such frequency bands and networks. Due to the usage of exceedingly heterogeneous networks, different communication scenarios, vast numbers of antennas, and broad bandwidths, the connection density will increase by

1,000 times. Because of satellites, unmanned aerial vehicles, and ultra-high-speed railways, mobility will be much higher than 500 km/h. When comparing its performance to that of other networks, several indicators such as network security, storage, range and coverage, cost efficiency and sustenance, and so on can be taken into account [7].

10.2 Evolution of 6G

The mobile communication sector has advanced dramatically, particularly in transmission technologies and frequency bands, from the first mobile system to the impending 6G mobile system. Each generation has its own unique traits, methods, and skills.

In the early 1980s, 1G cellular networks were introduced that relied on analog transmission for voice services. Nippon Telegraph and Telephone (NTT), a cellular system provider, started its business in Tokyo, Japan, in 1979. Then Europe launched the cellular system two years later. The most well-known analog systems were Total Access Communication Systems (TACS) and Nordic Mobile Telephones (NMT). The only problem with 1G networks was the use of analog signals for transmission such as: B. Security against poor quality calls, excessive power consumption and inadequate data capacity.

In 1991, the second-generation mobile network (2G) was introduced. Theoretically, this cellular network would include an integrated global distribution of multiple base stations (BS) that would allow users to connect multiple access points (FDMA, CDMA, and TDMA). Therefore, 2G technology can be used for compression/decompression methods such as 2G Global System for Mobile Communications (GSM), 2.5G General Packet Radio Service (GPRS), 2.75G Enhanced Data Rate for Global Evolution (EDGE) (codecs, etc.). Compared to 1G, 2G offers superior cellular services and digitally protected data transmission [8].

With the introduction of the third generation, the mobile network continued to evolve (3G). The network transitioned from a traditional mobile network to portable media devices as connection speeds increased (e.g., computers, gaming consoles, and tablets). CDMA2000, Wideband Code Division Multiple Access (WCDMA), and Time Division Synchronous Code Division Multiple Access are the three essential technologies for the 3G network (TD-SCDMA).

After that, 4G was deployed. Fourth generation is associated with the term "MAGIC," which stands for "mobile multimedia anywhere, global mobility solutions over integrated wireless and tailored services." Users can enjoy smooth network access and end-to-end IP transmission, as well as QoS management with greater service quality, mobility, and a data transfer rate of 20 Mbps.

After the original complete set of 5G specifications was established, the commercial application of 5G began in 2019. The advent of 5G signals the commencement of a global digital era with groundbreaking wireless technology standards, including data transfer rates, latency, mobility, and even the number of linked devices. The characteristics of 5G have truly distinguished this modern mobile network from its forerunners. 5G denotes the next step in the evolution of communication networks. Going beyond 5G to meet the expanding technical needs and demands at all levels is unavoidable, given the rapid technological breakthroughs and inventions over the last decade.

The first commercial 6G technology system is expected to be launched in 2030. By that year, a global digital civilization powered by improved and practically instantaneous wireless communication is envisioned. 6G is a self-governing computer that may replicate human intellect and consciousness and provides a variety of ways to communicate and interact with smart terminals (for example, through brain waves or neurological signals, eyes, fingers, and voice) [9].

As a vision for the future, and because 6G can use a very high spectrum compared to its predecessor, a 6G network with a multiband spectrum will spread hundreds of Gbit/s to Tbit. A connection to/s will be available. For example, this combination uses the 13 GHz band, the 30300 GHz band for millimeter waves, and the 0.0610 THz band for THz. The evolution of cellular networks from 2G to 5G is designed to serve people. That is, it reduces the delays caused by human response times such as visual response time (10 ms), auditory response time (100 ms), and perceptual response time (1 ms).

By the sixth generation, the wireless evolution from connected things to linked cognition is expected to be substantially altered. Furthermore, the delivery of ubiquitous AI services from the network's core to end devices necessitates 6G. Artificial intelligence (AI) will be crucial in the creation and optimization of 6G protocols, infrastructures, and operations, to put it another way. Below is a table that summarizes all the information about mobile networks throughout the years.

Network type	Year of launch	Functions/Application
1G mobile network	early 1980s	a) It used an analogue transmission signal that could only handle voice services at a speed of up to 2.4 kbps. b) Relied on analogue transmission for speech services. c) It also used a frequency range of 800–900 MHz, a bandwidth of 40 MHz, and a channel capacity of 30 kHz. d) Frequency division multiplexing was utilized in the first generation.
2G mobile network	1991	a) TDMA and CDMA were used in the second generation. b) 2G offered more services, such as "short message service" (SMS) and "multi-media service" (MMS), as well as higher service quality. c) 2G was enhanced to work in the 850–1900 MHz frequency band with a data rate of up to 64kbps.
3G mobile network	2003	a) High-quality internet access. b) Improved security by allowing users to connect to other wireless devices using user authentication capabilities. c) Defined by IMT-2000 technical specifications that include features of reliability and speed, namely a data transfer rate of at least 200 kbps.
4G mobile network	2009	a) Allowed users to connect to the network anytime and from any location. b) Allowed users to have smooth connectivity and improved service quality.
5G mobile network	2019	a) Users can expect considerable benefits from 5G, including data transfer rates of up to 10 Gbps, significantly lower latency (almost 10 ms) at larger capacity, reliability, and QoS. b) 5G is the first to use the mmWave band, a brand-new frequency band technology. c) Provides a single platform for a variety of applications, including improved mobile broadband communications, automated driving, virtual reality, and the Internet of Things.

(Continued)

(*Continued*)

Network type	Year of launch	Functions/Application
6G mobile network	2030 (expected)	a) 6G will play a key role in merging seamless wireless connectivity with numerous technology functions that support full-vertical applications. b) 6G will dramatically improve the data rate speed, up to 100–1000 times faster than 5G. c) In terms of capacity, 6G aims to sharply increase the capacity by up to 1000 times more than 5G. d) 6G will provide latency up to 10–100 μs.

10.3 Functionality

For faster speed and data rates, the radio frequency spectrum (RF spectrum) in 6G has been expanded to THz spectrum. THz waves with a frequency range of 0.1 to 10 THz and a wavelength of 30 to 3000 microns move through the spectrum. This allows for high transmission rates and broad broadband access, which might be useful in future mobile communication systems.

Furthermore, the THz band can keep up with the capacity of nanoscale cells up to micrometers across 10 m without sacrificing communication speed. Various systems, such as Holographic Beamforming, employ antenna arrays to send and receive a focused narrow beam with a very high gain. This is accomplished by focusing power in a restricted angular range. Beamforming increases the signal-to-interference-plus-noise ratio (SINR), which may be used to monitor individuals, while also improving coverage and throughput.

As a result, the 6G network's position accuracy improves dramatically. 6G Quantum communication is another technology that has to be utilized. According to quantum principles, data encoded in a quantum state (using photons or quantum particles) cannot be retrieved or replicated without changing the data (e.g., correlation of entangled particles and inalienable law).

In addition, the superposition property of qubits allows for a faster data transfer rate in QC. Quantum key distribution, quantum secret sharing, quantum teleportation, and quantum secure direct communication are now all possible with QC. One of the most important features of this technology is its capacity to greatly improve data security and reliability.

The design system and speed will be governed by edge intelligence. The main motive of 6G is to provide greater speeds using THz communications, which limits the communication range even further. As a result, in order to enable connection for objects in an environment, all devices must get services from a nearby access point (AP). On one side of the AP, there is a very high end and a lower band and speed end, and on the other side, there is a lower band and speed end. As a result, edge computing capability must be improved. Hence, 6G is expected to drive toward more robust edge capabilities to handle communication mismatches, as well as a new basic architecture to have more intelligent edge computing, thanks to the clever use of AI approaches.

Energy efficiency is another component of 6G that should be examined. The IRS (intelligent reflecting surface) is a revolutionary wireless communication idea. It has recently been seen as a potential technology capable of significantly lowering the energy consumption of wireless networks. Because it is extremely simple to install and uses very little energy, IRS-assisted communications could be utilised to dramatically improve the coverage of 6G networks. Indoor residual spraying (IRS) can be used to coat the external walls of buildings in the outdoor urban environment to reduce energy consumption. This allows for the expansion of 6G network coverage while simultaneously increasing energy efficiency.

Given the widespread use of the 5G mobile network, mobile devices' technological capabilities should be compatible with the upcoming 6G features.. Because some of the new 6G features are incompatible with 5G devices, increasing the technological capabilities of mobile devices for 6G may result in higher expenses. Individual devices, on the other hand, struggle to accommodate 1 Tbps speed.

10.3.1 Security and Privacy Issues

6G networks are giving high reliability, low latency and secure and efficient transmission services. However, most of these technologies come at a cost of new security and privacy concerns [1].

10.3.1.1 Artificial Intelligence (AI)

AI is commonly regarded as one of the major components of the future network architecture, when compared to all other technologies projected to be deployed in 6G networks [10]. To say that artificial intelligence has gotten a lot of attention in the subject of networking is an understatement. As a result of this increased focus, a growing number of new security and privacy issues have emerged. Although AI is apparently operated in isolated locations where huge amounts of training data and powerful but private processing hubs are available in the 5G network, AI will become a more central component of the 6G network. The physical layers, which comprise devices such as data cables and network infrastructure, and the computational layers, which include software-defined networks, network function virtualization, cloud/edge/fog computing, and so on, are the architectural levels that AI technologies serve.

Physical Layers
Many AI-based technologies, such as deep neural networks and supervised/unsupervised learning, can be applied to many physical layers. These approaches can not only increase physical layer performance by improving connectivity, but they can also forecast traffic and improve security. Unsupervised learning methods could be employed in the authentication process to improve the physical layer security. To prevent information breaches, machine learning–based antenna designs can be utilized in physical layer communication. Machine learning and quantum encryption algorithms might also be employed to defend the security of communication links in 6G networks, according to the researchers.

Network Architecture
It was believed that AI may improve edge security via security systems and fine-grained controls in terms of network architecture. It was also known that artificial intelligence (AI) technologies, specifically deep learning, might be utilized to detect dangers in edge computing. However, the concept has to be investigated further.

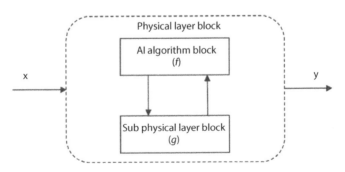

Figure 10.1 Physical layer in artificial intelligence [2].

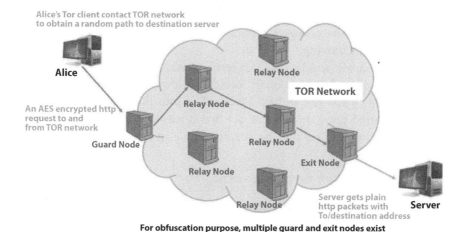

Figure 10.2 Network architecture for better security [2].

Other Functions

AI is useful in various areas besides the physical layers and network design, such as large data analysis, distributed AI, resource management, and network optimization (Figures 10.1 and 10.2). AI had previously been shown to aid in the detection of network anomalies and the provision of early warning measures to improve the security of 6G networks. It was also discovered that deploying distributed and federated AI in a 6G network eliminates the requirement for edge devices to communicate data, further enhancing network security. The impact of data correlation in various machine learning algorithms has been shown to result in an increase in privacy leaks.

10.3.1.2 Molecular Communication

A natural phenomenon observed among living beings with nanoscale structures is molecule communication. Microscale and nanoscale technologies are becoming a reality because of advances in nanotechnology, bioengineering, and synthetic biology over the last decade. Furthermore, the energy required for the formation and propagation of a molecular communication signal is negligible. Although this phenomenon has been researched for many years in biology, it has only recently been a research issue in the realm of communication. For 6G communications, molecular communication technology is a very promising technology. It is, however, a multidisciplinary technique that is still in its infancy. The core concept of molecular communication is the transmission of information via biochemical signals. It was demonstrated as a mobile molecular communication technique that allows the transmitter, receiver, and associated nodes to interact while moving.

Several security and privacy concerns relating to the communication, authentication, and encryption processes, however, have already been discovered. Only a few researchers have looked into the safety of molecular communication lines, despite the fact that it was known that this form of communication channel may be interrupted by an opponent.

10.3.1.3 Quantum Communication

Another communication technology with a lot of potential in 6G networks is quantum communication. One of its key advantages is that it may considerably improve data transfer security and dependability. The quantum state will be modified if an opponent listens in on, measures, or replicates anything in quantum communication. As a result, the recipient is unable to be ignorant of the interference. In theory, quantum communication might give complete security, and it could be ideal for long-distance communication with the right breakthroughs. It provides a slew of new features and raises communication to a level that older systems can't match. Quantum communication, on the other hand, is not yet a panacea for all security and privacy concerns. Although tremendous work has been made in establishing quantum cryptography for quantum communication, long-distance quantum communication remains a substantial barrier due to fiber attenuation and operation mistakes. To achieve entirely secure quantum communications, numerous different forms of quantum encryption and other approaches, such as quantum key distribution, quantum secret sharing, quantum secure direct transmission, quantum teleportation, and quantum dense coding, may be necessary. Furthermore, further details are needed on the security of quantum secure direct communication, which allows secret messages to be transmitted directly over a quantum channel without the use of a private key. There have also been cases where some quantum processes that use quantum key distribution models to protect key security have been discussed.

10.3.2 Blockchain

In a 6G network, Blockchain technology offers a lot of potential applications. Network decentralization, distributed ledger systems, and spectrum sharing are among examples. Network decentralization based on blockchain technology has the potential to improve network administration and performance. The same can be said for the usage of blockchain in distributed ledger technologies, which would greatly improve authentication security. In fact, blockchain has the potential to be one of the most disruptive Internet of Things technologies. Furthermore, by incorporating blockchain technology into a spectrum sharing scheme, issues such as low spectrum utilization and spectrum monopoly could be addressed while also ensuring spectrum use.

Access control, authentication, and communication mechanisms all play a role in blockchain security and privacy. There is a blockchain radio access network design that can protect and manage network access and authentication among trustless network elements. There is also a new conceptual architecture for mobile service authorization based on blockchain technology. It was previously known that a means of utilizing the blockchain to improve the security of media access protocols and cognitive radio in order to acquire access to unlicensed spectrums existed. Furthermore, despite the fact that the 6G network's decentralized architecture means that a hacker can only change records if more than 51% of the nodes are under his control (indicating that it is secure enough), there is no trusted third party responsible for secure data storage and management when security breaches occur. The hash capacity necessary to validate transactions in a blockchain-based network was discovered to have a negative impact on security.

10.3.3 TeraHertz Technology

Despite their widespread use in 5G networks, mm-wave bands are insufficient in the 6G environment due to the necessity for high transmission rates. In any event, the Radio Frequency (RF) band is nearly full, and future technology cannot use it. Terahertz technology has accelerated as a result of these considerations. The 0.1–10 THz range, which has more spectrum resources than the mm-wave frequency, is used for terahertz communication. It also makes use of both electromagnetic and light waves. There are several advantages to adopting the THz band. To begin with, THz communication technology may be capable of supporting data rates of up to 100 Gbps. Second, eavesdropping would be restricted due to the narrow beam and short pulse duration employed in THz communication, resulting in increased communications security. Third, THz waves have a very low attenuation through specific materials, which means they could be used in a wide range of applications. Furthermore, THz communication transmission can be extremely directed, reducing intercell influence dramatically.

The energy consumption of THz communication has been identified as a significant issue. The size of 6G cells must be reduced from "small" to "tiny," necessitating the development of more complicated hardware and designs. THz, like all other technology, has its own set of security and privacy concerns. The majority of these are concerned with authentication and malicious conduct. Concepts such as the electromagnetic signature of THz frequencies, for example, could be employed in physical layer authentication methods. Furthermore, while THz communication is usually thought to make eavesdropping harder, a signal broadcast via narrow beams could still be intercepted by an eavesdropper. They do, however, talk on how to defend against such an eavesdropping attack.

10.3.4 Visible Light Communication (VLC)

The employment of visible light communication technology to address the growing need for wireless connectivity is a viable option. VLC has been investigated for several years and has already been applied in a variety of applications, including indoor positioning systems and the Vehicular Ad Hoc Network (VANET) network. VLC offers greater bandwidths and can resist electromagnetic interference better than RF, which has interference and significant latency. The advancement of VLC technology has also been aided by the advent of solid-state lighting. Some researchers have sought to employ LEDs for high-speed data transfer since LEDs can switch between different light intensities at a very fast rate. LiFi is a VLC system that supports multiple access and has the potential to provide high-speed services to a large number of mobile users. However, some flaws in VLC technology are impeding its advancement. Because intense natural light will impact transmissions, the main application cases for VLC should be indoors. Malicious behaviors and communication mechanisms are among the security and privacy concerns raised by VLC. It was discovered that an attacker must be in the line of sight of the victim in order to initiate an assault on a running VLC operation. Obviously, this would make it easier for the attacker to be detected. There are a number of approaches and protocols that can be used to secure security transmission data, including SeeVLC, a preceding solution for VLC links that improves the physical layer's security, and others. Furthermore, it was discovered that eavesdroppers' assistance can decrease the security of VLC technology.

Comparison of communication performance indicators between 5G and 6G.

Communication performance indicators	Definition	5G requirements	6G requirements
Peak data rate	Maximum data rate can reach per user/device under ideal conditions	20 Gbps	>100 Gbps
User experience data rate	Data rate achievable for users/mobile devices in target coverage of	0.1-1 Gbps	>1 Gbps
Communication latency	Time interval between sending packets at source and receiving them at destination	1 ms	0.1 ms
Area traffic capacity	Total traffic provisioned per geographic unit	10 Mbit/s/m^2	1 Gbit/s/m^2
Connection density	Total number of connected and/or accessible devices out of units of area	1/m^2	10-100/m^2
Mobility	Maximum relative speed between transmitter and receiver when certain QoS is met	500 km/h	1000 km/h
Reliability	Probability that will successfully transmit a fixed size packet within the specified maximum time	0.99999	0.9999999
Timing accuracy	Precision time synchronization between devices	Microsecond level	Nanosecond level

Comparison of 5G services and 6G services.

Service type		5G services	6G services
Communication	Basic telecom business	Basic telecom services, VoNR, new voice, 5G messaging, etc.	XR, holographic telepresence, multi-sensory interconnection, etc.
	Data connection	On-demand mobile data connectivity	Higher-performance on demand mobile data connectivity
Information		UE positioning, some information	Provide basic information services natively, including wireless sensing, better network information distribution, and public sector information.
Computing		MEC	Support convergent computing services natively, such as processing speed, storage, AI, etc.

10.4 6G Security Architectural Requirements

Since 6G is intended to be a more open network than 5G, the boundaries between on-grid and off-grid become increasingly blurry. As a result, current network security measures such as IPsec and firewalls are not strong enough to protect a network from external intruders [3]. The 6G security architecture must support the essential zero trust (ZT) security concept in mobile networks to mitigate this issue. ZT is a security paradigm that prioritizes the protection of system resources.

The following lines describe the security requirements that the 6G network security architecture must manage and handle.

Virtualization Security Solution: Virtualization protection issues want the usage of a machine with a steady virtualization layer, which incorporates a protection software/technology that identifies dangerous hidden software, along with rootkits. The hypervisor must be able to segregate storage and network services that use secure protocols such as TLS, SSH, and VPN into entirely separate categories. Virtual Machine Introspection (VMI) is a hypervisor capability that examines and identifies security concerns for each virtual machine (VM) by analyzing vCPU register information, file I/O, and communication packets.

Automated Management System: The most important thing to deal with open-source security issues is to control the vulnerabilities created by using, updating and removing open-source software. Therefore, rapid threat detection requires an automated management system capable of detecting vulnerabilities and applying updates. Another step is needed to ensure that patched software is installed quickly and securely using a secure OTA method. In addition, a security governance framework should be created to manage (1) long-term open-source vulnerabilities, (2) changes in developer opinion, and (3) deployment of solutions security.

Data security using AI: To ensure that AI systems are protected against AML, they must be public about how they protect users and mobile communication systems. The first step of the process is to create AI models in a trusted system. In addition, a technique such as a digital signature should be used for authentication if the AI models running in the user equipment (UE), radio access network (RAN) and core have been updated or modified maliciously by a hostile attack. When a system detects an unsafe AI pattern, it must perform self-healing or recovery operations. The system should also limit data collection to AI training to trusted network elements.

Preserving Users' Privacy: In order to maintain the confidentiality of the user's personal information, the user's personal information must be stored and used according to the protocols agreed between the service provider, the mobile network operator (MNO), subscriber and MNO. The 6G system protects personal information in a Trusted Execution Environment (TEE) and trusted software, and reduces or anonymizes the amount of data that is made public when it is used. Before the MNO publishes personal information, its authenticity and permission must be confirmed. When it comes to user information, another alternative is to use homomorphic encryption (HE) so that the data can be accessed in encrypted form. To protect user location and usage privacy, AI-based solutions, such as learning-based privacy-aware offloading systems, can be implemented.

Post-Quantum Cryptography: Current asymmetric key encryption methods should be abandoned in 6G systems because quantum computers would make them insecure. Many scholars have focused on post-quantum cryptography (PQC) solutions such as network-based cryptography, cypher-based cryptography, multivariate polynomial cryptography, and hash-based signatures. From 2022 to 2024, the US National Institute of Standards and Technology (NIST) will select the best PQC algorithms as part of its PQC research. The key length currently under discussion for PQC should be several times the length of Rivest-Shamir-Adleman (RSA). PQC is expected to be computationally more expensive than the current RSA approach. Therefore, PQC must be integrated in accordance with the HW/SW service and performance requirements of the 6G network.

6G Security Challenges
This section discusses some of the challenges related to AI/ML in the 6G system.

- Trustworthiness: As AI manages cybersecurity, the stability of machine learning models and components becomes important.
- Visibility: Real-time monitoring of security operations based on AI and machine learning to ensure controllability and reliability.
- Ethical and legal aspects: Some customers or applications may be limited by AI-based optimization strategies. Who is responsible for the failure of AI-controlled security services? Are AI-powered security solutions uniform in protecting all users? Who is responsible for the failure of AI-controlled security services? [4]
- Extensibility and viability: Secure data exchange is necessary to protect the privacy of federation learners. The required compute, network, and storage resources need to be scalable, which is an AI/ML hurdle.
- Controlled security tasks: Using AI/ML security solutions in combination with large-scale data operations can incur significant overhead [5].

In the process of learning and inference, the flexibility of the model must be safe and flexible.

Sense performance indicators for two system configurations.

System parameters		System configuration 1	System configuration 2
	Central frequency	6 GHz	30 GHz
	Bandwidth	400 MHz	2 GHz
	Number of antenna elements/ element gain	256/8 dBi	512/8 dBi
	Transmitting power of BS	55 dBm	40 dBm
	Inter-site distance	500 m	200 m
	Reference RCS	0.1 m2	0.1 m2
	Target maximum velocity	120 km/h	120 km/h
	Coherent processing interval	5 ms	1 ms
Sensing performance at cell edge	Distance resolution	0.375 m	0.075 m
	Velocity resolution	5 m/s	5 m/s
	Angular resolution (azimuth/ zenith)	7.2°/7.2°	3.6°/7.2°
	Distance accuracy	~0.1 m	~0.1 m
	Velocity accuracy	~1 m/s	~7 m/s
	Angular accuracy (azimuth/ zenith)	~2°/2°	~5°/10°

Typical use cases and application scenarios for ISAC

Category of ISAC use cases	Use cases	Application scenarios
Coarse-grained sensing	Monitoring the weather and the quality of the air	Agriculture, meteorology, and daily living services
	Traffic flow detection, pedestrian volume statistics, and intrusion detection	Smart transportation, security surveillance
	Localization, tracking, and measurement of the target object's range, speed, and angle	Radar application scenarios
	Mapping the environment	For car and UAV (Unmanned Aerial Vehicle) navigation, smart driving and city
Fine-grained sensing	Face, motion, and position recognition	Interactive intelligence, gaming, and smart homes
	Monitoring of vital indicators (heartbeat, respiration, etc.)	Medical and health care
	Imaging, detecting materials, and analyzing composition	Industry, biomedicine, and security inspection

10.5 Future Enhancements

1. Since DLT and 6G are expected to work together, vulnerabilities in blockchain and smart contracts can have an unintended impact on 6G networks. Of course, when implementing a DLT/blockchain solution on a 6G network, users should always adhere to the available procedures to mitigate the above security threats. However, the implementation of certain security techniques may be more important to the public blockchain than to the private blockchain. For example, smart contracts are adopted by every node in the

blockchain network, so debugging and modifying smart contracts can be a time-consuming task. Smart contracts are very important in DLT/blockchain systems to enable automation, so it is important to make sure they are accurate. In addition, smart contracts need to be properly validated for correct functionality before deploying to hundreds of blockchain nodes.

2. Quantum computing is likely to become commercially available in the coming years and poses a significant threat to current cryptography. Quantum computing is currently intended for use in 6G communication networks to detect, mitigate, and prevent security gaps. Scientists have already begun exploring quantum-resistant technologies and cryptographic solutions to prepare for the threat posed by quantum computing in the future 6G era. Lattice-based, code-based, hash-based, and multivariate-based cryptography is a part of post-quantum cryptography primitives. The grid calculation problem works better with IoT devices in the current environment. The short key length makes it suitable for 32-bit architectures.

3. AI and machine learning are expected to play a major role in 6G. AI and ML, on the other hand, make 6G intelligence network management systems vulnerable to AI/ML-related threats. There are various AI/ML methods to counter these dangers. To improve resilience, enemy training inserts modified instances into the training data, similar to attacks. Another defense method is defensive distillation. This is the output of a previously trained network and is based on the idea of transferring knowledge from one neural network to another via soft labels that reflect the possibilities of different classes. These are used for training instead of hard labels that assign all data to a particular class. Both solutions have been successful against both evasive and enemy attacks.

4. Unauthorized receivers can intercept signals transmitted in line of sight (LOS) even with very narrow beams. As a result, THz communications can lead to attacks on data transmission, eavesdropping, and access control. There is evidence that unauthorized users can intercept the signal by placing an object on the transmission path and scattering radiation in that direction. It has been proposed to characterize channel backscatter and detect some, if not all, eavesdroppers.

6G efficiency indicators

6G efficiency indicators	Communication indicators	Sensing indicators	Computing indicators
Spectral efficiency	Definition: throughput provided per cell per frequency resource 6G requirements: 2-3 times higher than 5G	The time and frequency resources required to complete one sensing task	N/A
Energy efficiency	Definition: number of bits that can be transmitted per unit of energy, or the amount of energy required to transmit 1 bit 6G requirements: More than 100 times improvement in network energy efficiency compared to 5G. 10 to 100 times improvement in terminal energy efficiency compared to 5G.	Energy required to complete one sensing task	Number of operations available per unit of energy
Cost efficiency	Definition: the number of bits that can be transmitted per unit cost, or the cost required to transmit 1 bit. 6G requirements: more than 100 times improvement compared to 5G	Cost to complete one sensing task	Number of operations available per unit cost

10.6 Summary

We are currently in the early stages of the 5G commercialization process, which is expected to usher in a significant revolution—or, at the very least, evolution—in the mobile wireless communications industry. The revolutionary component of 5G is how it "extends" previously limited use cases, such as mobile internet, to include ultra-reliable low-latency communications and enormous machine-type connections. As a result, the mobile communications industry invented the phrase "5G triangle." Although 5G has only recently been deployed in several countries and areas and is still in its early stages, researchers and developers are actively working on the creation of 6G, particularly in relation to new communication techniques and technologies. The sixth generation is an upcoming mobile system that would dramatically alter future mobile wireless networks. It is projected to be deployed in 2030. The "5G triangle" will grow into the "6G hexagon" when more dimensions are added to open new industrial use cases, according to this analysis. Furthermore, 6G is expected to deliver a high data rate and ultra-low latency of up to 10 Tbps and 10–100 s, respectively. It will also improve spectrum efficiency and connection density by a factor of 10–100 over 5G. Furthermore, 6G will enable intelligent IoE and the goal of everything being connected 100 percent intelligently, bringing new applications beyond IoE. The sixth generation ushers in a new era of seamless machine-to-human interactions, object intelligence, and the merging of the virtual and physical realms. The core pillars of the 6G vision are ultra-high reliability, ultra-high flexibility, ultra-high privacy and security, and ubiquitous coverage.

References

1. https://www.analyticsvidhya.com/blog/2018/07/using-power-deep-learning-cyber-security/
2. Kim, H. (2022). AI-Enabled Physical Layer. In: Artificial Intelligence for 6G. Springer, Cham. https://doi.org/10.1007/978-3-030-95041-5_8
3. Siriwardhana, Yushan & Porambage, Pawani & Liyanage, Madhusanka & Ylianttila, Mika. (2021). AI and 6G Security: Opportunities and Challenges. 10.1109/EuCNC/6GSummit51104.2021.9482503.
4. Wang, Minghao & Zhu, Tianqing & Zhang, Tao & Zhang, Jun & Yu, Shui & Zhou, Wanlei. (2020). Security and privacy in 6G networks: New areas and new challenges. Digital Communications and Networks. 6. 10.1016/j.dcan.2020.07.003.

5. Abdel Hakeem, Shaimaa & Hussein, Hanan & Kim, Hyungwon. (2022). Security Requirements and Challenges of 6G Technologies and Applications. Sensors. 22. 1969. 10.3390/s22051969.
6. P. Porambage, G. Gür, D. P. M. Osorio, M. Liyanage, and M. Ylianttila, "6G security challenges and potential solutions," in Proc. IEEE Joint Eur. Conf. Netw. Commun. (EuCNC) 6G Summit, 2021, pp. 1–6.
7. M. Ylianttila et al., "6G white paper: Research challenges for trust, security and privacy," 2020. [Online]. Available: arXiv:2004.11665.
8. C. Castro. (2020). Korea Lays Out Plan to Become the First Country to Launch 6G. [Online]. Available: https://www.6gworld. com/exclusives/korea-lays-out-plan-to-become-the-first-country-tolaunch-6g/
9. M. Wang, T. Zhu, T. Zhang, J. Zhang, S. Yu, and W. Zhou, "Security and privacy in 6G networks: New areas and new challenges," Digit. Commun. Netw., vol. 6, no. 3, pp. 281–291, 2020.
10. Y. Siriwardhana, P. Porambage, M. Liyanage, and M. Ylianttila, "AI and 6G security: Opportunities and challenges," in Proc. IEEE Joint Eur. Conf. Netw. Commun. (EuCNC) 6G Summit, 2021, pp. 1–6.

11
A Trust-Based Information Forwarding Mechanism for IoT Systems

Geetanjali Rathee[1*], Hemraj Saini[2], R. Maheswar[3] and M. Akila[4]

[1]*Department of CSE, Netaji Subhas University of Technology, Dwarka Sector-3, New Delhi, India*
[2]*Department of CSE, School of Computing, DIT University, Dehradun, Uttarakhand, India*
[3]*Department of ECE, Centre for IoT & AI (CITI), KPR Institute of Engineering and Technology, Coimbatore, India*
[4]*Department of CSE, KPR Institute of Engineering and Technology, Coimbatore, India*

Abstract

The intelligent information gathering, analysis and monitoring process reduces the human efforts in almost every field of applications such as e-healthcare, industry manufacturing, e-voting and intelligent transportation systems, etc. However, the involvement of smart devices/systems for processing and taking intelligent decisions invites a number of severe security attacks in the network. The intruders may change the communication pattern and perform various changes inside the network for their own benefits. The aim of this paper is to propose an efficient and trust-based information forwarding mechanism for the IoT systems. The proposed approach provides the security while transferring, collecting, analyzing or taking accurate decisions by the smart devices in an appropriate manner. The proposed approach is validated against traditional mechanism over various security metrics.

Keywords: Trust-based scheme, information forwarding, intelligent systems, secure data transmission, secure IoT

Corresponding author: geetanjali.rathee123@gmail.com

11.1 Introduction

In recent years, the demand for intelligent and smart-based systems has been increasing at a very high pace. Organizations are adopting every new or modern technology that is arriving in the market for improving their industrial growth [1, 2]. IoT-based applications such as smart industrial manufacturing, smart home automation, intelligent transportation mechanism, and e-healthcare systems are a few ongoing research areas where IoT systems are replacing human interactions while taking shipping/manufacturing decisions, gathering/analyzing patient records, or intelligent switch handling of home equipment [3, 4]. Though there are a number of advantages to adopting or using such automation systems, intelligent/smart systems, they are still not being fully adopted by all the organizations in the market. The most severe and drastic issue is security, where organizations are afraid that an intruder or third party in the network will attack, steal or control their confidential information [5, 6].

11.1.1 Need of Security

Whenever an intelligent automation system is adopted by any organization in the market, it's the responsibility of each organization to ensure and provide 100% confidentiality and security to its client/user. Any breach in security may drastically affect the security of the system that may further lose the trust of its clients [7, 8]. Figure 11.1 presents the intelligent communication process by any organization for intelligent data forwarding or collection mechanism by the system where a number of industrial IoT devices are taking independent decisions and accurate actions upon requirement by its environment.

Various security schemes have been proposed by several scientists or authors; however, trust-based methods are the most general and effective security scheme or approach to provide an efficient communication process in the network [9, 10]. The communication process done by highly trusted nodes may further improve the system performance and its overall throughput that may directly increase/improve the growth of any organization.

11.1.2 Role of Trust-Based Mechanism in IoT Systems

The IoT systems play a significant role in ensuring security against various malwares, and intruders while distributing the information among various

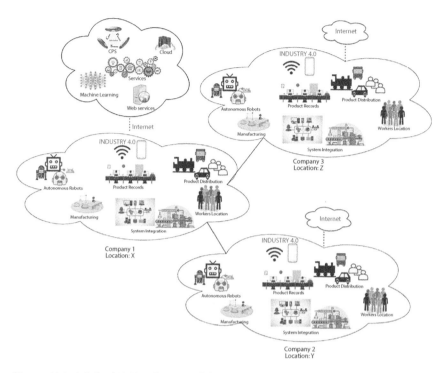

Figure 11.1 A hybrid IIoT architecture [9].

devices. After discussing the importance of smart devices and their need of security while communicating among each other, in this subsection we will discuss various security techniques using trust-based and blockchain mechanisms in the network. A number of trust-based systems have been proposed by various scientists and authors such as direct trust, indirect trust, hybrid trust, etc. [11]. In addition, the blockchain mechanism is used to ensure transparency while providing security during sharing of information among smart devices.

The architecture of trust-based and blockchain network is presented in Figure 11.2 that represents the trust computation of IoT devices associated with each device. The trust of each device is computed using various techniques such as direct, indirect and hybrid mechanisms. The trust of each device is stored in a database along with blockchain mechanism. In addition, the blockchain mechanism is integrated with trust-based devices to ensure transparency among devices [12, 13]. The role of blockchain and trust-based intelligent devices is used in various applications such as industry 4.0, smart homes, smart city, intelligent transportation systems, etc.

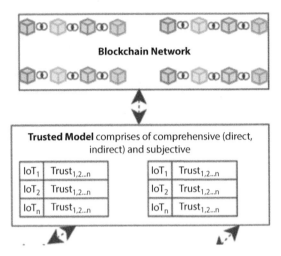

Figure 11.2 The architecture of trust-based and blockchain network [26].

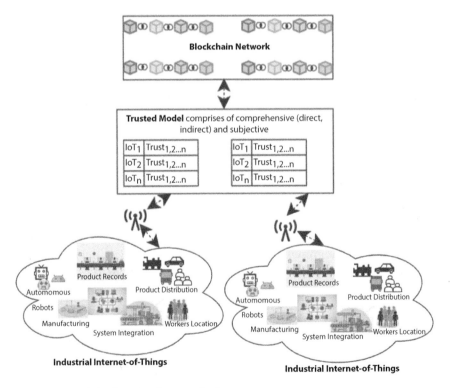

Figure 11.3 A hybrid IIoT architecture [26].

The text below explains industry 4.0 along with blockchain and trust-based systems.

Figure 11.3 presents the role of trust-based and blockchain mechanisms in smart devices. In this we have taken the example of industrial Internet of Things where a number of communications and distributions are accomplished through intelligent devices. Here the manufacturing, recording of products, manufacturing, autonomous robots, system integration and product distribution is done through intelligent devices. In addition, the security of intelligent devices is ensured using trust-based mechanism where the trust of each device is computed before distributing the information. Further, the blockchain mechanism is integrated with trust-based mechanism in order to ensure transparency among the devices [14, 15].

11.1.3 Contribution

The aim of this paper is to propose an efficient and secure trust-based mechanism that forwards the ongoing data/information to its neighboring nodes or in the networking environment after computing their trust values. A trust value of each node will be computing based upon their behavior that will further decide the acceptance and blockage of communication process in the network [16]. In addition, the blockchain mechanism is associated with trust-based systems in order to further ensure the transparency in the network. The proposed work is validated and implemented against various security metrics against the traditional approach used by other scientists such as accuracy, data alteration rate, blockchain network of legitimate devices. All the parameters are discussed with graphs and proper explanation while sharing the information among devices.

The remaining organization of the paper is as follows. Section 11.2 describes the literature survey and related work proposed by various authors/academicians to ensure a secure communication process. Section 11.3 presents the in-depth detail of proposed trust-based information forwarding communication mechanism. The implementation and validation are discussed in section 11.4. Finally, section 11.5 concludes the paper along with some future directions.

11.2 Related Works

This section discusses the various security methods such as authentication, key management, encryption, blockchain based and trust-based schemes proposed by various researchers. Table 11.1 depicts the comparison of

Table 11.1 Literature survey.

Author	Technique	Measuring parameter	Limitation
Sindhuja [17]	Secure routing mechanism	Sensor hub clustering.	Suffers from communication delay.
Abed [18]	Web of things, IoT and Blockchain network	Importance of blockchain and IoT for ensuring efficient communication in the network.	Energy consumption while tracking the mechanism again and again.
Rathee et al. [19]	Trusted and Blockchain mechanism for IIoT	Proposed trusted evaluation mechanism for ensuring a secure and transparent communication process upon integration of blockchain technology in the network.	Authentication process leads to delay.
Rajavel et al. [20]	Trust-aware pricing for VIoT	Proposed a trust-based resource and pricing allocation mechanism with the aim of maximizing the profits in the network.	Issues during dynamic behavior of the network.
Rathee et al. [21]	Trusted Mechanism using blockchain in Industry 4.0	The proposed scheme verified against various security metrics such as attack strength, message alteration and false authentication.	Need to consider accuracy.

(*Continued*)

Table 11.1 Literature survey. (*Continued*)

Author	Technique	Measuring parameter	Limitation
Rathee *et al.* [22]	Secure Vaccine Distribution	The authors proposed an ANN scheme to provide viable and blockchain mechanism for distributing the vaccine in the network.	Needed to analyze the computation delay of trust.
Lashmi and Pillai [23]	Decision-making approach	The authors projected home-based mechanism by focusing on the face recognition and anomaly detection schemes to ensure a secure communication in the network.	Need to monitor continuously for ensuring a secure communication in the network.
Lee *et al.* [24]	Trusted framework based on ontology	The proposed mechanism is analyzed by determining the highest degree of trust based on trusted ontology and estimated the degree of trust for each communicating device.	Authors have not considered the real time scenario while communicating among each other.

already proposed schemes/methods along with their measuring environments and limitations in the various applications.

Sindhuja [17] proposed an efficient and secure routing mechanism by selecting the cluster heads for congestion-free IoT systems. The author used sub-strategy and load adjusting and proficiency approach while ensuring security in the network. Abed [18] discussed the wireless network, IoT and blockchain importance while ensuring an efficient and effective communication in the network. The author highlighted the importance of the web of things along with IoT devices and blockchain technology together in the network. The author pointed out some of the open issues for the research community for ensuring a better information transmission process in the environment. Rathee *et al.* [19] proposed a trusted and blockchain-based cyber securing mechanism against IIoT in the network. The authors have a trusted evaluation mechanism for ensuring a secure and transparent communication process upon integration of blockchain technology in the network. They verified the proposed approach over various security measures.

Rajavel *et al.* [20] proposed a trust-aware pricing scheme for Vehicular IoT in the network. The authors proposed a MobiTrust mechanism for identifying various types of evidences in the network. Further, they proposed a trust-based resource and pricing allocation mechanism with the aim of maximizing the profits in the network. The proposed mechanism is simulated over various benchmarks.

Lashmi and Pillai [23] projected a decision mechanism approach for detecting the security in which intruders may compromise a number of devices in the network. The authors projected home-based mechanism by focusing on the face recognition and anomaly detection schemes to ensure a secure communication in the network. They further validated the proposed mechanism by determining various illegal activities of the intruders by regularly monitoring and capturing the IoT devices' communication.

Lee *et al.* [24] projected a trusted framework based on ontology of the individuals according to their perspectives and purposes. The proposed mechanism is analyzed by determining the highest degree of trust based on trusted ontology and estimated the degree of trust for each communicating device.

A number of secure communication and information transmission schemes/methods have been proposed by various scientists/researchers; however, the trust-based security where computation and communication determine the legitimacy of each communicating device is still at its early stage. Further, the proposed schemes are not fully adopted for IoT devices according to today's era.

Though various security schemes have been proposed by several authors, the proposed mechanisms have their own limitations and drawbacks that lead to a research question on the security of IoT-based applications. This paper proposes a trust-based information forwarding scheme for IoT-based systems along with their validation and implementation steps in the subsequent sections. In addition, the proposed scheme is validated and verified against a number of security metrics against existing and traditional mechanisms.

11.3 Estimated Trusted Model

Trust is considered one of the significant performance metrics in order to ensure or analyze the legitimacy of each communicating device in the network. A number of trust-based schemes have been proposed and used by various researchers and scientists for computing the trust of each device, such as direct trust, indirect trust, hybrid trust, estimated trust, etc. Each trust-based approach has its own significance while choosing the communication pattern and transmitting of information scenario. Among them in this paper, we have used estimated trust value where the communicated devices trust depends on forwarding information to their neighboring devices. The estimated trust value analyses the legitimacy by depending upon three different factors such as total trust, processed trust and estimated trust. The neighboring devices will estimate a trust value depending upon their communication behavior or transmission of information among each other [25].

The designing of a trusted information mechanism is crucial for any application having IoT devices in the network. The system architecture of a proposed model is illustrated in Figure 11.1, having several types of data forwarding IoT-devices/nodes in the network. The trust computation of IoT device mainly consists of three parts as presented in eq. (11.1), where T_{total} represents the total computed trust and T_e, T_p, and T_t denotes the trust computation for estimated, processed and transmitting information respectively.

$$T_{total} = T_e + T_p + T_t \qquad (11.1)$$

In addition, T_n can be defined as the trust computation for neighboring device, T_{amp} is the amplifying trusted value estimated by the nodes i, and d is defined as the transmission distance.

$$\begin{cases} T_e = n \times T_n \\ T_e = n \times (T_n + T_{amp}d) \end{cases} \qquad (11.2)$$

Therefore, the overall trust computation is further defined as:

$$T_{total} = (T_n + T_{amp}d) + T_p \qquad (11.3)$$

Further, in order to accurately recover the original information by the neighboring nodes, the gathered trust can be further defined as the basis vector over sparse basis and basis vector as:

$$X_{N\times 1}^T = \sum_{i=1}^{N} \alpha_i \phi_i = \phi_N \times N^\alpha \qquad (11.4)$$

Where, $X = x_1, x_2, \ldots x_n$, α_i, $i = (1,2, \ldots N)$ and $\alpha = \alpha_1, \alpha_2, \ldots \alpha_N$

11.4 Blockchain Network

Recently, a number of wireless communication devices are present in the network for the ease of the users. In various applications such as healthcare systems, smart cities, smart homes, industrial internet of things, a large number of intelligent devices are used to determine the communication in the network. The smart devices further need proper privacy and security while maintaining the communication and transmission of information. Till now, a number of security protocols and algorithms have been proposed by various researchers/scientists; however, a number of cyber issues and security issues are still present in the network. Blockchain technology has been considered as one of the recent paradigms in today's era for ensuring a secure and transparent communication in the network. Only a few organizations have adopted this technology and it is still in its early stage. Further, in order to ensure security and transparency in the network, the blockchain network is maintained. The blockchain consists of all those nodes that are trusted and being surveilled by the network for ensuring transparency during communication. The integration of trust mechanism along with blockchain network of trusted nodes can be discussed in Algorithm 1. The presented algorithm 1 illustrates the trusted computation and blockchain mechanism for ensuring a secure communication in the network. The estimated trust approach is used to analyze the legitimacy of each communicating device. The estimated value of each device ensures security and permits further transmission of information in the network. In addition, the blockchain mechanism is used to maintain the continuous surveillance in the network for better protection of the devices.

TRUSTED MECHANISM FOR IOT SYSTEMS 249

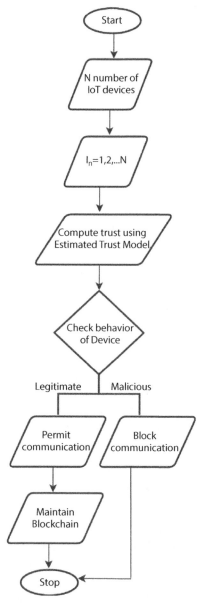

Figure 11.4 Flowchart of proposed model.

Algorithm 1
Begin
Step 1: *Input*: Number of IoT devices
Output: Device is ideal or fraud.
Given: Trust-based computation and a blockchain network.
Step 2: a) Establish the networking environment
For all nodes $I_n = \{I=1, 2,....N\}$
For p=1 to N then
Compute Trust of each device using estimated trust model
If (Device is ideal) then
Permit further communication
Else
Block/deny further communication
End if
b) Maintain a Blockchain network of highly trusted devices.
Step 3: Each ideal device is surveillance using blockchain
End For
End For

The flowchart of the same is presented in Figure 11.4.

The aim of this paper is to propose a secure and efficient trust-based communication process among IoT devices in the network. In order to conduct further surveillance, a blockchain network is maintained for keeping the record of highly trusted nodes that are providing the services in the network.

11.5 Performance Analysis

The simulation of the proposed framework is validated against Rajavel *et al.* (also known as baseline approach) where the authors proposed a cryptographic and key management scheme for ensuring secure communication among smart devices in the network. The authors proposed a trust-aware pricing scheme for Vehicular IoT in the network. The authors also proposed a MobiTrust mechanism for identifying various types of evidences in the network. Further, they proposed a trust-based resource and pricing allocation mechanism with the aim of maximizing the profits in the network. The proposed mechanism is simulated over various benchmarks.

The performance analysis of the proposed mechanism is validated against Wu *et al.* over various security metrics such as accuracy, and data alteration results. The number of devices is considered as 50 at present for analyzing the behavior of each device during transmission of information in the network. The proposed and existing mechanism is analyzed over a small area of the network to check the accuracy and competence of trust computation along with its complexity in the network. The data alteration and accuracy metrics can be easily evaluated over a small part of the network for determining the behavior of each communicating device. The proposed mechanism can be further analyzed over a large area of the network, whereupon increasing the size of the network, the devices may attain sustainability and understand the dynamic pattern of information transmission in the network. The performance measurement while analyzing the various metrics may further affect the nature of communicating nodes in the network.

11.5.1 Dataset Description and Simulation Settings

The proposed framework is tested over a synthesized dataset having a number of legitimate devices upon establishment of network. The devices are further altered intentionally for the purpose of showing the outperformance and accuracy of the proposed framework over existing scheme. The simulation is done over MATLAB having 700 × 700 area size with 50 number of nodes. The nodes are altered at the rate 10% upon increasing the network size from 5 to 50 count. Each node can start or transmit the communication process and information in the network at any time.

11.5.2 Comparison Methods and Evaluation Metrics

The proposed mechanism is compared against the following schemes:

Accuracy: The proposed and existing approaches are measured against accuracy metrics that determine the number of times both approaches are successfully able to detect or identify the number of altered or legitimate number of nodes in the network.
Data alteration rate: This metric is used to determine the alteration rate that can be successfully done by any intruder while transmitting or forwarding the messages in the network.
Blockchain network: It is used to measure the block size and time required to validate a particular node during communication or before adding in to the blockchain network.

The above-said evaluation metrics are analyzed against various security metrics over the existing mechanism and the reason of the proposed approach outperformance with proper graphs appears in the results discussion section below.

11.6 Results Discussion

This section describes the results and discusses the proposed mechanism by explaining various graphs against accuracy, data alteration rate, blockchain of legitimate devices. Figure 11.5 depicts the accuracy graph of proposed and existing approaches while increasing the number of altered nodes in a network size of 5 to 50 devices. The proposed mechanism is successfully able to outperform the existing mechanism because of its trust computation. The computed trust value ensures the legitimate nodes involvement while performing the communication or forwarding the information to its neighboring nodes.

The computed trust value does not include overhead of any external memory and delay while ensuring the security in the network. Figure 11.6 presents the data alteration rate that represents the number of nodes that can be successfully altered by the intruders upon establishment of the network. The proposed approach outperforms in this case as the number of nodes whose trust values is higher can take part in the communication process while the nodes having lower trust value could never be a part of communication process. The data rate alteration because of computed trust rates further improves the data alteration rate in comparison to the existing approach.

Finally, Figure 11.7 represents the blockchain of legitimate devices in the network. In order to surveille the entire network and provide security without increasing the overhead or delay may further include the

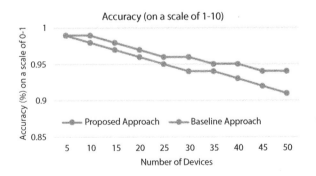

Figure 11.5 Accuracy.

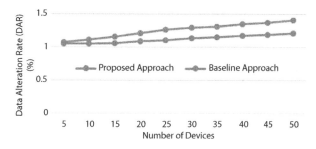

Figure 11.6 Data alteration rate.

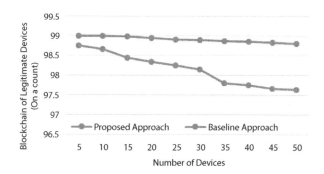

Figure 11.7 Blockchain of legitimate devices.

blockchain process. The legitimate nodes blockchain network further ensures transparency and provides higher-level security during the communication process in the network. However, in the existing mechanism, it generates lots of communication and computational overheads along with increased delays in the network.

11.7 Empirical Analysis

Further, the empirical analysis may also determine the overall performance of the system by identification of their computation overhead, delays and security concerns while analyzing the dynamic patterns of communicating devices in the network. The proposed and existing mechanisms are again analyzed using an empirical analysis where a number of metrics are considered for further analysis of both approaches. The number of empirical factors is considered as below:

Computation overhead: the computation overhead where each device needs to do some of the computation for identifying the legitimacy of each device. The computations can be done at the end of communicating device itself or devices may communicate among each other for further analysis of the network.

Response delay: in cases where devices are analyzed by communicating among each other, the delay may also be high in that case. The communication among devices while analyzing or identifying the legitimacy of each communicating device may result in large response delays in the network that may further invite a number of security threats in the network.

Security concerns: the dynamic communication in the network may invite a number of networking and communicating security hacks in the network. Security can be further compromised where intruders may easily understand the patterns of communicating devices and mimic exactly legitimate device in the network.

This empirical study is further analyzed over proposed and existing scenarios where the proposed mechanism outperforms existing methods because of less overhead trusted algorithms. The estimated trust degree computation may easily recognize the legitimacy of the device without having huge computations or calculations in the network. In addition, the response delay and other security hacks can be easily traced using blockchain-based technology. Each and every communicating device is placed on blockchain for their continuous surveillance and tracing of each device in the network.

The time complexity of the proposed phenomenon is big O because of sensor nodes usage and their battery life while transmitting the information in the network. Each and every device is communicating among each other while transmitting the information in the network. Since the proposed phenomenon opted a forwarding mechanism where the trust value of each device is decided based estimated trust scheme where the neighboring devices trust values are involved to finalize the legitimacy of any current device. Each device took some amount of time to analyze the final decision of any communicating device to allow or block that from the network. In a network of size 'N', the trust value of 'N' different devices will be computed after communicating among each other. Therefore, the overall complexity of the algorithm (as mentioned above in Algorithm 1) is Big (O) notation or simply defined as the time taken by each device for executing the trust that is involved in transmitting of information in the network.

11.8 Conclusion

This paper has proposed a secure and trusted communication procedure for forwarding the information among nodes in the network using IoT devices. The proposed systems compute the trust of each communicating node according to their processed, estimated and transmitted ratio that can be further decided for allowing or blocking the remaining communication in the network. The proposed mechanism is validated and implemented against various security protocols in the network against a traditional approach. The depicted graphs and results outperform the proposed approach because of continuous surveillance and trust value computation before permitting them for communication in the network. The dynamic pattern recognition along with their security threats can be further discussed as the future direction of this paper.

References

1. Elrawy, M. F., Awad, A. I., & Hamed, H. F. (2018). Intrusion detection systems for IoT-based smart environments: a survey. *Journal of Cloud Computing*, 7(1), 1-20.
2. Liao, H. J., Lin, C. H. R., Lin, Y. C., & Tung, K. Y. (2013). Intrusion detection system: A comprehensive review. *Journal of Network and Computer Applications*, 36(1), 16-24.
3. Pradhan, B., Bhattacharyya, S., & Pal, K. (2021). IoT-based applications in healthcare devices. *Journal of Healthcare Engineering*, 2021.
4. Khraisat, A., Gondal, I., Vamplew, P., & Kamruzzaman, J. (2019). Survey of intrusion detection systems: techniques, datasets and challenges. *Cybersecurity*, 2(1), 1-22.
5. Hajjaji, Y., Boulila, W., Farah, I. R., Romdhani, I., & Hussain, A. (2021). Big data and IoT-based applications in smart environments: A systematic review. *Computer Science Review*, 39, 100318.
6. Butun, I., Morgera, S. D., & Sankar, R. (2013). A survey of intrusion detection systems in wireless sensor networks. *IEEE Communications Surveys & Tutorials*, 16(1), 266-282.
7. Rathee, G., Ahmad, F., Kurugollu, F., Azad, M. A., Iqbal, R., & Imran, M. (2020). CRT-BIoV: a cognitive radio technique for blockchain-enabled internet of vehicles. *IEEE Transactions on Intelligent Transportation Systems*, 22(7), 4005-4015.
8. Khanna, N., & Sachdeva, M. (2019). Study of trust-based mechanism and its component model in MANET: Current research state, issues, and future

recommendation. *International Journal of Communication Systems*, 32(12), e4012.
9. Rathee, G., Sharma, A., Kumar, R., & Iqbal, R. (2019). A secure communicating things network framework for industrial IoT using blockchain technology. *Ad Hoc Networks*, 94, 101933.
10. Khan, Z. A., & Herrmann, P. (2017, March). A trust based distributed intrusion detection mechanism for internet of things. In *2017 IEEE 31st International Conference on Advanced Information Networking and Applications (AINA)* (pp. 1169-1176). IEEE.
11. Muzammal, S. M., Murugesan, R. K., & Jhanjhi, N. Z. (2020). A comprehensive review on secure routing in internet of things: Mitigation methods and trust-based approaches. *IEEE Internet of Things Journal*, 8(6), 4186-4210.
12. Calvaresi, D., Leis, M., Dubovitskaya, A., Schegg, R., & Schumacher, M. (2019). Trust in tourism via blockchain technology: results from a systematic review. *Information and Communication Technologies in Tourism* 2019, 304-317.
13. Hasselgren, A., Rensaa, J. A. H., Kralevska, K., Gligoroski, D., & Faxvaag, A. (2021). Blockchain for increased trust in virtual health care: proof-of-concept study. *Journal of Medical Internet Research*, 23(7), e28496.
14. Montecchi, M., Plangger, K., & Etter, M. (2019). It's real, trust me! Establishing supply chain provenance using blockchain. *Business Horizons*, 62(3), 283-293.
15. Singh, M., & Kim, S. (2018, February). Trust bit: Reward-based intelligent vehicle commination using blockchain paper. In *2018 IEEE 4th World Forum on Internet of Things (WF-IoT)* (pp. 62-67). IEEE.
16. Anjum, A., Sporny, M., & Sill, A. (2017). Blockchain standards for compliance and trust. *IEEE Cloud Computing*, 4(4), 84-90.
17. Sindhuja, M. (2022, March). Secure and Efficient Cluster Head Selection and Routing in WSN for IoT through Congestion Free Mechanism. In *2022 6th International Conference on Computing Methodologies and Communication (ICCMC)* (pp. 271-275). IEEE.
18. Abed, A. A. (2021, December). IoT, WoT, and Blockchain Technologies. In *2021 2nd Information Technology to Enhance E-learning and Other Application (IT-ELA)* (pp. 201-201). IEEE.
19. G. Rathee, C. A. Kerrache and M. Lahby, TrustBlkSys: A Trusted and Blockchained Cybersecure System for IIoT, in *IEEE Transactions on Industrial Informatics*, 2022, doi: 10.1109/TII.2022.3182984.
20. Rajavel, D., Chakraborty, A., & Misra, S. (2022). MobiTrust: Trust-Aware Pricing Scheme for Edge-Based Mobile Sensor-Cloud for Vehicular IoT. *IEEE Transactions on Vehicular Technology*.
21. Rathee, G., Ahmad, F., Jaglan, N., & Konstantinou, C. (2022). A Secure and Trusted Mechanism for Industrial IoT Network using Blockchain. arXiv preprint arXiv:2206.03419.

22. Rathee, G., Garg, S., Kaddoum, G., & Jayakody, D. N. K. (2021). An IoT-Based secure vaccine distribution system through a blockchain network. *IEEE Internet of Things Magazine*, 4(2), 10-15.
23. K. Lashmi, A. S. Pillai, Ambient intelligence and iot based decision support system for intruder detection, in: *2019 IEEE International Conference on Electrical, Computer and Communication Technologies (ICECCT)*, IEEE, 2019, pp. 1-4.
24. L. Zhang, G. Feng, S. Qin, Y. Sun, B. Cao, Access control for ambient backscatter enhanced wireless internet of things, *IEEE Transactions on Wireless Communications*, 21 (7), July 2022, 5614-5628.
25. Wu, D., Yang, B., Wang, H., Wu, D., & Wang, R. (2016). An energy-efficient data forwarding strategy for heterogeneous WBANs. *IEEE Access*, 4, 7251-7261.
26. Rathee, G., Kerrache, C. A., & Lahby, M. (2022). TrustBlkSys: A trusted and blockchained cybersecure system for IIoT. *IEEE Transactions on Industrial Informatics*.

About the Editors

S. Sountharrajan, PhD, is an associate professor in the School of Computing Science and Engineering at VIT Bhopal University. He is also the Division Head specializing in artificial intelligence and machine learning. He has published numerous papers in scientific journals and conferences and has delivered 32 guest lectures in reputed universities and institutions.

R. Maheswar, PhD, is an associate professor at the School of Electrical & Electronics Engineering, VIT Bhopal University. He has over 18 years of teaching experience at various levels and has published 60 papers at scientific journals and conferences.

Geetanjali Rathee, PhD, is an assistant professor in the Department of Computer Science and Engineering at Netaji Subhas University of Technology, New Delhi. She has six patents to her credit and has published over 40 papers in scientific journals and 15 papers at scholarly conferences. She has published book chapters, as well, and one book. She is a regular reviewer of various reputed journals.

 **M. Akila** is a professor in computer science. He has delivered 32 lectures at various universities, seminars, and conferences, and he has three patents to his credit. He is deeply involved in industrial and entrepreneurial activities.

Index

3G, 57, 219,221
4G, 57, 219,221
5G, 18, 57, 165, 169, 174, 217, 218, 219, 220, 221, 222, 223, 224, 227, 229, 230, 236, 237
6G, 140, 217, 218-238

Analysis of Various Security and Assaults, 122
Attacks in Data Link Layer, 67
Attacks in Physical Layer
   eavesdropping, 67
   jamming, 67
   tampering, 67
   side channel attack, 67
   sybil attack, 67
   random inference, 67
   timing attack, 67
Wireless Medical Sensor Network (WMSN), 71
A Biometric-Based, 1, 5, 11, 16
Adam Enhancer, 38
Advanced Secured Effective Encryption Algorithm (ASEEA), 112, 117, 119
ARP Spoofing, 154
Artificial Intelligence, 53, 105, 133, 149, 151, 160, 166, 168, 172-175, 213, 217, 218, 220, 223, 224, 237
Attacks in Application Layer, 68
Attacks in Network Layer, 68
Attacks in Transport Layer, 68

Bitcoin, 25, 73, 130, 131, 132, 136, 142, 143, 145-147
Black-box, 149, 151, 159, 172
Blockchain, 18, 127-139, 141-147, 201, 204-209, 211, 213, 217, 226, 227, 234, 235, 241-257
Blockchain Classification, 135
Blockchain Components and Operation
   data, 139
   hash, 139
   MD5, 139
   SHA 256, 140
Blockchain Revolution Drivers, 133
Blockchain Technology, 130-132, 135, 142, 144, 226, 246, 248
Boltzmann Machine, 22, 23, 30, 34, 51, 182
Botnet, 92, 94, 156

Categorisation, 35, 37, 39, 40, 41
CDH (Customized Diffie-Hellman), 112, 113, 109
Character sequence vectors, 190, 191, 192
Checksum, 139, 140
Clustering, 9, 10, 16, 106, 107
Convolutional Neural Networks, 23, 27, 29
Cookie session, 155
Cryptography, 106, 107, 122, 140, 226

262 INDEX

Cyber Security, 27, 29, 53, 54, 55, 103, 128, 165, 174, 196, 197
Cyberattacks, 22, 90, 149, 152, 165, 166, 167, 168, 169, 170, 171
Cyberattacks, 90, 92, 152, 165-171
Cyber insurance, 73

DALEX, 174, 178
Data integrity, 66, 68, 139
Data Preprocessing, 35
Deep Belief Network, 21, 23, 30
Deep Learning Algorithms for Malware Detection, 26
Deep MD5 matching, 182
Deep Neural Networks, 22, 29, 182, 190
DeepNet Algorithm, 35, 36
Denial of Services, 81
Denial-of-service , 3, 17, 99, 171
Depuration, 4, 18
Diffie-Hellman, 106, 107, 109, 112, 113, 115, 119, 120,
Distributed Denial of Service, 64, 79, 84, 87
Distributed Ledger (DL) Technology, 217
DL model, 37
DNS Spoofing, 154
Dynamic trusted approach, 199

ELI5, 160, 163
Elliptic Curve Cryptography-Advanced Encryption Standards, 109
Ethereum, 136, 139, 145, 147
Experimental Analysis, 12
Explainable Artificial Intelligence (XAI), 149, 151, 160
F1-Score, 170, 192
Firewall, 73, 75, 82, 85, 217, 230
Fuzzy membership, 1, 5, 11, 16

Gated Recurrent Unit, 27
Graph Neural Network Explainer, 159

Hash Function, 4, 16, 138, 139, 140
Healthcare, 17, 55, 56, 58, 60, 61, 71, 72, 73, 77, 78, 101, 103, 127, 129, 131, 133, 142, 147, 151, 159, 166, 174, 204, 239, 240, 248, 255
Holographic Connectivity, 217, 218
HTTPS Spoofing, 154

Information Forwarding, 239, 246, 247
Internal Behavior (IBij), 206
Internet Control Message Protocol Attacks, 83
Intrusion Detection System, 68, 83, 93, 100, 101, 186, 269

LEACH, 107, 110, 112, 117, 118
LiFi, 46, 84, 144, 228
LIME (Local Interpretable Model agnostic Explanations), 158-160, 162-164, 170, 174, 220
Long short- term memory (LSTM), 30, 182

MaleVis Dataset, 22, 40, 42, 43, 44, 45, 46, 48, 50
Malicia dataset gatherings, 47
Malicious Traffic Detection, 89, 90, 94
Malimg Dataset, 22, 32, 34, 40, 43, 44, 46, 48, 49, 50
Malware Attack
 ransomware, 22, 26, 65, 72-74, 78, 95, 155, 169
 spyware, 26, 155, 176
 botnet, 92, 94, 156
 fileless malware, 156
Malware Datasets
 android malware dataset, 28
Man-in-the-Middle (MITM) Attack, 154
Man-in-the-Middle Attack, 66, 109, 123, 135, 152

Index

MD5, 139, 140, 141, 147, 182
Metaheuristic Methods for Malware Detection, 25
Military, 55, 57, 60, 61
ML Methods
  k-nearest neighbor, 44
  logistic regression, 44, 160, 183
  naïve bayes, 44, 52, 182, 206
  SVM, 37, 42, 44, 45, 54, 181, 183, 184
  decision tree, 44, 182, 183
  random forest, 44, 54, 181, 182
  adaboost, 25, 26, 42, 44, 45
Modified Rivest Shamir Adleman, 109
Molecular communication, 217, 225, 226
Multilayer perceptron, 175, 184

Natural Language Processing (NLP), 23, 159, 160, 196
Nodes, 1, 2, 3, 9, 11, 13, 14, 15, 16, 68, 106, 107, 118, 136, 207, 211, 248, 252

Optical Wireless Communication (OWC), 218
Optimizers
  gradient descent (GD), 179, 180
  mini-batch GD, 179, 180
  stochastic GD,, 179, 180
  SGD-momentum, 179, 180
  adadelta, 179, 180
  root mean square (RMS), 179, 180
  adagrad, 179, 180
  adam optimizer, 175, 179, 180, 181, 184

Pharming, 153
phishDiff, 181
Phishing Attack, 63, 64, 152, 153, 154, 175-178, 181, 183, 184, 195-197
Phishing Attack, 153
Post-Hoc, 158

Post-Quantum Cryptography, 217, 231, 235
Privacy Scheme, 199
Proposed Methodology, 35, 112, 181, 194
Pseudorandomness, 7, 8, 9

Quantum key distribution, 222
Quantum secret sharing, 222
Quantum secure direct communication, 222
Quantum teleportation, 222

Random Forest (RF), 44, 181
Recurrent Neural Networks, 23, 27, 30, 179, 182, 185, 187
Replay attacks, 77, 80, 82, 181
Restricted Boltzmann Machine, 22, 23, 30
RUS Boost, 22, 172, 181

SDN Secure Control and Records Plane, 86
Security Risks in Healthcare, 72
Self-organizing maps (SOMs), 183
Sensor Node, 1, 2, 3, 5, 7, 68, 106
Session Hijacking, 155
SHA 256, 139, 140, 141
SHAP (Shapley additive Explanations), 159-162, 170, 173, 217
Signal-to-interference-plus-noise ratio, 222
Simda class, 48
Smart Banking, 179, 180
Smishing, 153
SMOTE Boost, 181
SoftMax, 22-24, 36, 39
Softmax Classifier, 23, 36, 39, 51
Software Defined Networking, 93, 94, 115, 117
SOREL-20M Dataset, 28
Spear phishing, 153
Spectrum sharing, 222, 226, 227
SQL Injection, 66, 69, 152, 156

SSL handshakes, 140
Stacked Auto Encoders, 21, 31
Stock Market, 142
Storj, 139
Support Vector Machine (SVM), 54, 181
System Architecture, 34, 107, 247

Telecast Transmission, 11
TensorFlow, 40, 54
Third-party regulators, 142
Three-Factor Authentication, 3, 18
Traffic Analysis, 62, 63, 76, 79
Traffic Collection, 89, 90
Trails, 13, 35

Uniform Resource Locator, 177, 186, 187
UUID, 140

Visible Light Communication (VLC), 217, 228

Wavelet transform, 9
Web page Phishing Detection Dataset, 186-188, 195
Whaling, 153
Wi-Fi Eavesdropping, 55, 154
Wireless Sensor Network, 1, 2, 4, 17, 18, 19, 57, 59, 77, 78, 102, 105, 106, 107, 124, 125, 126, 127, 215, 255

XAI, 149, 151, 152, 157-160, 165-174
XAI and Its Categorization
   intrinsic or Post-Hoc, 158
   model-specific or model-agnostic, 159
   local or global, 159
   explanation output, 159

Zero Trust, 217, 230
Zero-Day Exploit, 152, 156

Also of Interest

Other Books in the Series

WEARABLE AND NEURONIC ANTENNAS FOR MEDICAL AND WIRELESS APPLICATIONS, Edited by Arun Kumar, Manoj Gupta, Mahmoud A. Albreem, Dac-Binh Ha, and Er. Mohit Kumar Sharma, ISBN: 9781119791805. This new volume in this exciting new series, written and edited by a group of international experts in the field, covers the latest advances and challenges in wearable and neuronic antennas for medical and wireless applications.

NEXT-GENERATION ANTENNAS: Advances and Challenges, Edited by Prashant Ranjan, Dharmendra Kumar Jhariya, Manoj Gupta, Krishna Kumar, and Pradeep Kumar, ISBN: 9781119791867. The first book in this exciting new series, written and edited by a group of international experts in the field, this exciting new volume covers the latest advances and challenges in the next generation of antennas.

Check out these other Related Titles

RESOURCE MANAGEMENT IN ADVANCED WIRELESS NETWORKS, Edited by A. Suresh, J. Ramkumar J; M. Baskar and Ali Kashif Bashir, ISBN: 9781119827498. Written and edited by a team of experts in the field, this exciting new volumes provides a comprehensive exploration of cutting-edge technologies and trends in managing resources in advanced wireless networks.

RF CIRCUITS FOR 5G APPLICATIONS: Designing with mm Wave Circuitry, Edited by Sangeeta Singh, Rajeev Kumar Arya, B.C. Sahana and Ajay Kumar Vyas, ISBN: 9781119791928. This book addresses FinFET-based analog IC designing for fifth generation (5G) communication networks and highlights the latest advances, problems, and challenges while presenting the latest research results in the field of mmwave integrated circuits design.

WIRELESS COMMUNICATION SECURITY: Mobile and Network Security Protocols, Edited by Manju Khari, Manisha Bharti, and M. Niranjanamurthy, ISBN: 9781119777144. Presenting the concepts and advances of wireless communication security, this volume, written and edited by a global team of experts, also goes into the practical applications for the engineer, student, and other industry professionals.

ADVANCES IN DATA SCIENCE AND ANALYTICS, edited by M. Niranjanamurthy, Hemant Kumar Gianey, and Amir H. Gandomi, ISBN: 9781119791881. Presenting the concepts and advances of data science and analytics, this volume, written and edited by a global team of experts, also goes into the practical applications that can be utilized across multiple disciplines and industries, for both the engineer and the student, focusing on machining learning, big data, business intelligence, and analytics.

ARTIFICIAL INTELLIGENCE AND DATA MINING IN SECURITY FRAMEWORKS, Edited by Neeraj Bhargava, Ritu Bhargava, Pramod Singh Rathore, and Rashmi Agrawal, ISBN 9781119760405. Written and edited by a team of experts in the field, this outstanding new volume offers solutions to the problems of security, outlining the concepts behind allowing computers to learn from experience and understand the world in terms of a hierarchy of concepts.

MACHINE LEARNING AND DATA SCIENCE: Fundamentals and Applications, Edited by Prateek Agrawal, Charu Gupta, Anand Sharma, Vishu Madaan, and Nisheeth Joshi, ISBN: 9781119775614. Written and edited by a team of experts in the field, this collection of papers reflects the most up-to-date and comprehensive current state of machine learning and data science for industry, government, and academia.

MEDICAL IMAGING, Edited by H. S. Sanjay, and M. Niranjanamurthy, ISBN: 9781119785392. Written and edited by a team of experts in the field, this is the most comprehensive and up-to-date study of and reference for the practical applications of medical imaging for engineers, scientists, students, and medical professionals.

SECURITY ISSUES AND PRIVACY CONCERNS IN INDUSTRY 4.0 APPLICATIONS, Edited by Shibin David, R. S. Anand, V. Jeyakrishnan, and M. Niranjanamurthy, ISBN: 9781119775621. Written and edited by a team of international experts, this is the most comprehensive and up-to-date coverage of the security and privacy issues surrounding Industry 4.0 applications, a must-have for any library.

CYBER SECURITY AND DIGITAL FORENSICS: Challenges and Future Trends, Edited by Mangesh M. Ghonge, Sabyasachi Pramanik, Ramchandra Mangrulkar, and Dac-Nhuong Le, ISBN: 9781119795636. Written and edited by a team of world renowned experts in the field, this groundbreaking new volume covers key technical topics and gives readers a comprehensive understanding of the latest research findings in cyber security and digital forensics.

DEEP LEARNING APPROACHES TO CLOUD SECURITY, edited by Pramod Singh Rathore, Vishal Dutt, Rashmi Agrawal, Satya Murthy Sasubilli, and Srinivasa Rao Swarna, ISBN 9781119760528. Covering one of the most important subjects to our society today, this editorial team delves into solutions taken from evolving deep learning approaches, solutions allow computers to learn from experience and understand the world in terms of a hierarchy of concepts.

MACHINE LEARNING TECHNIQUES AND ANALYTICS FOR CLOUD SECURITY, Edited by Rajdeep Chakraborty, Anupam Ghosh and Jyotsna Kumar Mandal, ISBN: 9781119762256. This book covers new methods, surveys, case studies, and policy with almost all machine learning techniques and analytics for cloud security solutions.

SECURITY DESIGNS FOR THE CLOUD, IOT AND SOCIAL NETWORKING, Edited by Dac-Nhuong Le, Chintin Bhatt and Mani Madhukar, ISBN: 9781119592266. The book provides cutting-edge research that delivers insights into the tools, opportunities, novel strategies, techniques, and challenges for handling security issues in cloud computing, Internet of Things and social networking.

DESIGN AND ANALYSIS OF SECURITY PROTOCOLS FOR COMMUNICATION, Edited by Dinesh Goyal, S. Balamurugan, Sheng-Lung Peng and O.P. Verma, ISBN: 9781119555643. The book combines analysis and comparison of various security protocols such as HTTP, SMTP, RTP, RTCP, FTP, UDP for mobile or multimedia streaming security protocol.

Printed and bound by CPI Group (UK) Ltd, Croydon, CR0 4YY
20/11/2023

08192117-0002